A Clash of Empires

Library of Modern Middle East Studies

Volume

1 Water in the Middle East: Legal, political and commercial implications edited by *J. A. Allan* and *Chibli Mallat*
ISBN 1 85043 645 2

2 Yemen and the U.S.A. from 1962 *Ahmad Almadhagi*
ISBN 1 85043 772 6

3 Workers and the Working Class in the Ottoman Empire and the Turkish Republic 1839–1850 edited by *Donald Quataert* and *Erik J. Zürcher*
ISBN 1 85043 875 7

4 From Land Reform to Revolution: The political economy of agricultural development in Iran *Fatemeh E. Moghadam*
ISBN 1 86064 006 0

5 Iranian Intellectuals in the Twentieth Century *Ali Gheissari*
ISBN 1 85043 968 0

6 Middle Eastern Politics and Ideas: A history from within edited by *Moshe Maoz* and *Ilan Pappé*
ISBN 1 86064 012 5

7 Secret War in the Middle East: The covert struggle for Syria, 1949–1961 *Andrew Rathmell*
ISBN 1 85043 992 3

8 Nationalism, Minorities and Diasporas: Identity and rights in the Middle East edited by *Kirsten E. Schultze, Martin Stokes* and *Colm Campbell*
ISBN 1 86064 052 4

9 Water, Peace and the Middle East: Negotiating resources in the Jordan Basin *J. A. Allan*
ISBN 1 86064 055 9

10 Islamic Politics in Palestine *Beverley Milton-Edwards*
ISBN 1 86064 049 4

11 'Independent Iraq': The monarchy and British influence, 1941–58 *Matthew Elliot*
ISBN 1 85043 729 7

12 The Army and the Creation of the Pahlavi State in Iran, 1910–1926 *Stephanie Cronin*
ISBN 1 86064 105 9

13 A Clash of Empires: Turkey between Russian Bolshevism and British Imperialism, 1918–1923 *Bülent Gökay*
ISBN 1 86064 117 2

A Clash of Empires

Turkey between Russian Bolshevism and British Imperialism, 1918–1923

BÜLENT GÖKAY

Tauris Academic Studies
LONDON • NEW YORK

Published in 1997 by Tauris Academic Studies
an imprint of I.B.Tauris & Co Ltd
Victoria House
Bloomsbury Square
London WC1B 4DZ

A full CIP record for this book is available from the British Library

A full CIP record for this book is available from the Library of Congress

ISBN 1 86064 117 2

Set in Monotype Garamond by Philip Armstrong, Sheffield

Printed and bound in Great Britain by WBC Ltd, Bridgend, Mid Glamorgan

Contents

Maps

Acknowledgements

I am especially indebted to the supervisor of my PhD thesis at the University of Cambridge, Jonathan Haslam, for his guidance, encouragement and friendship. I would like to thank Richard Langhorne whose insightful criticism and moral support helped me to start my research with a clear sense of direction. I wish to record the kind assistance of Michael Franklin, Wolfson College, Cambridge, who helped me out over various problems. I would also like to thank Ben Fowkes who, through his teaching, led me to appreciate the value of critical thinking.

I owe special thanks to my dear friends, Uygur Kocabasoglu of the Middle East Technical University, Ankara, and Kate Fleet, curator of the Skilliter Center for Ottoman Studies, for those long, stimulating and inspiring discussions.

I also wish to thank the staff of the following institutions whose assistance greatly facilitated my research: The University Library, Churchill College Archive Centre, and Trinity College Library, Cambridge; the Public Record Office, the Library of the School of Oriental and African Studies, and Marx Memorial Library, in London; the Baykov Library, University of Birmingham; the Lenin Library, the Central Party Archives, and the Archives of Foreign Ministry, in Moscow. I am especially indebted to Oleg V. Naimov for facilitating my speedy admission to the Central Party Archives, and the late A. A. Bykov of the Foreign Ministry Archives for granting me special access to some documents which were originally put aside for reclassification.

I benefited enormously from the expertise of Stephen White who gave me good advice on how to prepare the text for publication. I owe a great debt to Louise J. Wyne-Williams who painstakingly read the manuscript and spent many hours correcting, clarifying, and helping to rework the text. A special mention must be reserved for Anna Enayat at I.B. Tauris for her helpfulness, efficiency, and her dynamic editorial style. I should also like to thank Philip Armstrong who drew the maps.

I wish to send my grateful thanks to my friend Gorgun Taner who never disappointed me with his exceptional skills in finding obscure books and for sending large parcels from Turkey.

And finally I should like to record my heartfelt gratitude to Berrak to whom I will always be indebted for her companionship, tolerance and encouragement.

Introduction

This book documents the emergence of the Turkish Republic from the ruins of the Ottoman Empire at the end of the First World War. The war of 1914–18 was the first general war, involving all of the major powers of the day, and most of the minor. Every aspect of international relations was permanently and profoundly affected by the war which touched almost every corner of the world. Among the many consequences of the conflict was the fall of four great imperial dynasties – the Hohenzollerns, the Habsburgs, the Romanovs and the Ottomans – and with them their empires.

The Ottoman Empire, so long in decay, officially died with the Mudros Armistice in October, 1918. Under the terms of the armistice the lands of the Empire were clearly placed in the hands of the British and their allies. It was envisaged that various Allied-controlled enclaves would replace the defeated Empire and it seemed unlikely that there would be any resistance to this scheme.

Britain signed the Armistice in the name of the Allied Powers – demonstrating its leadership in the immediate post-war Turkish settlement. From a military, strategic, and political point of view, Turkey was of exceptional interest to Britain. Great Britain was a major world power responsible for crucial commitments in the eastern Mediterranean, Mesopotamia and, most importantly, India. It could not allow Turkish affairs to be settled without its direct and active participation. It naturally assumed the leading role and the greatest commitment in the region among the Allied powers.

Britain emerged from the war with its imperial domain much greater than before. It acquired new territories in the Middle and Near East[1] and in former German colonies in Africa and the Pacific.[2] The favourable circumstances of victory and post-war euphoria created an overly optimistic view of building new regional orders in the areas that were added to the British Empire. The intention was to consolidate the war gains and to maintain a strong British influence at the expense of the defeated parties. In the wake of their dramatic defeat, Turkey and Germany seemed no longer to be contenders. Nor was Russia perceived as a player in the imperial game due to the direct consequences of post-revolutionary political chaos.

At the time of the armistice with the Ottoman Empire, the former Ottoman colonies in Mesopotamia[3], Palestine, Syria and the other Arab provinces had already been occupied by the Allied troops and because of this their eventual separation from the Empire was already thought of as a matter of fact. What was left of the former lands of the Ottoman Empire was more or less limited to those areas in Anatolia and Eastern Thrace where the majority of the population was Turkish.

The situation was complicated by an unexpected development in the last year of the war. The Ottoman armies had made a show of force and used the opportunity to occupy the Caucasus. This meant that the Turkish armistice also put the Caucasian lands under the control of the British troops. At a time when the former strong power of the region had been toppled by the Bolshevik Revolution, the British Empire emerged as the only force that could dictate its terms over an area populated by 25 million people across almost 400,000 square miles.

It did not take long for this optimistic picture to dissolve. One can identify two primary factors accounting for the failure of the earlier British vision: the emergence and swift growth of a Turkish national resistance movement in Anatolia, and strong Bolshevik claims to legitimacy in the Caucasus. Faced with a pragmatic alignment between these two forces, Britain was forced to recognise the fact that it could not afford the manpower and financial resources needed to control the region.

The Turkish–Soviet co-operation, consolidated first in the Caucasus, succeeded in counterbalancing British influence in the region. British sponsorship of the Greek army in Anatolia and the Allied support of the anti-Bolshevik Russian armies in Russia drove the Turkish nationalists and the Bolsheviks into each other's arms.

The Historical Setting

The post-war conflict in the Near East found Britain and Soviet Russia lined up on opposite sides, Moscow supporting the Turks, London the Greeks. Throughout the nineteenth century the imperial conflict between Russia and Britain in the Near and Middle East, the territory encompassed by the Ottoman Empire, formed a branch of what is known as the 'Eastern Question'. The 'Eastern Question' itself was the issue of what was to happen to the vast territories from the Balkans in the west to the Caucasus in the east, and from the Black Sea in the north to the Persian Gulf in the south when the 'sick man of Europe'(namely the declining Ottoman Empire) eventually ceased to exist.[4]

The emergence of Turkey as an important factor in the dynamics of Anglo–Russian relations was closely connected to the strategic location of the Turkish Straits. Consisting of the Dardanelles, the Sea of Marmara, and the Bosphorus, the Turkish Straits lie at the junction of the key strategic and maritime routes which served for centuries to link the commercial activities of the West to the natural resources of the East.

The deepening imperial conflicts were reflected in the attitude of the powers to the status of the Straits, where both had interests at stake which they believed vital to their state. Control of maritime transport and military command of the Straits would, it was believed, lead to the domination of the Ottoman Empire, a prize which neither Britain nor Russia would concede to the other. On the other hand, the Straits involved part of or access to the major commercial and sea-routes of both states and this represented a strategic and logistical problem.[5]

Traditionally Russia had endeavoured to achieve two goals on its southern flank: to prevent any hostile attack from the rear through the Black Sea and to keep open its only exit to warm waters. From a military standpoint the Straits was the Achilles' heel for the Tsarist Russia as free access to the Black Sea would enable other powers to attack its southern shore. On the other hand, the Straits were a key to the empire's economic well-being, being the commercial outlet for the natural reserves of the Caucasus and the grain of the Ukraine.

Equally, for Britain, the supreme power in the Mediterranean, Russian access to the Straits would threaten its position in the eastern Mediterranean, the Middle and Near East. Russian control in the Straits was also considered harmful to the safety of British lines of communication with its Asian empire, in particular India. In turn, the reward for Britain was egress to the north. The Black Sea was considered the most effective point where British naval supremacy could be brought to play against Russia. That is why the British had always sought the right to pass through the Turkish Straits if and when they required.

It was no coincidence that for most of the eighteenth and nineteenth centuries both powers had been seeking essentially the same goal. Each would have liked an agreement with the Ottoman Empire that would secure free passage for its ships through the Straits and closure for those of its opponents. Hence the concept of a *mare clausum* emerged as the best solution to this imperial rivalry. This meant closure of the Straits to ships of war in time of peace with the maintenance of the Ottoman ruler as the 'guardian of the Straits'. As might be expected, the rule was sometimes violated, depending upon the particular stage of international rivalry and the extent of external pressure upon the Ottoman Sultan when, from time to time, he summoned the assistance of 'friendly and allied powers'.[6] During the eighteenth and nineteenth centuries there were repeated confrontations between the two powers over the Black Sea and the Straits which followed a similar pattern with Russia attempting to open the Straits to Russian warships and Britain taking measures to counteract the Russian threat.

The competing policies of Russia and Britain during the nineteenth century were not always related to this issue. Russia's steady advance in Central Asia and its involvement in Afghan affairs constituted the major conflict in Anglo–Russian relations in the second half of the nineteenth century. However, since the most effective way that Britain could counter Russian superiority in ground troops was still the threat of a naval attack in the Black Sea, the status of the Straits remained a major issue in British foreign policy.

At the turn of the century, rivalry in the region lost much of its intensity. The British government at this time tended more and more to base Britain's defence of its Near Eastern interests on its position in Egypt. Constantinople[7] and the Straits, on which two generations of Britain's foreign policy had been focused, now played a secondary role for the political and military leaders of Britain. Russia for its part was to look for nearly a decade to the Far East, to Manchuria, Korea, and north China, as a natural area of its expansion. The opportunities open to it in the Near East seemed limited and difficult to exploit effectively.

Although the Near East was apparently relegated to a secondary role in the political rivalry of the great powers, regional economic competition intensified. Indeed, the region saw a considerable increase of the economic activity of the great powers and of economic rivalry between them. In the two decades before 1914 the Ottoman Empire was not the most important area of international economic rivalry, but it was certainly one of the most manifest and most glamorous. Above all, the activities of Germany caused increasing anxiety to both the British and Russian governments who tried to secure concessions from the Ottoman government to safeguard their positions.

German economic influence had been increasing in the Near East since the 1890s. Germany began to search for new markets in order to provide for the demands of its growing industries. The building of the Baghdad Railway (Baghdadbahn) seemed to symbolise the growth of its influence. The line was designed to connect Haydar Pasha, one of the Asiatic suburbs of Constantinople, with Basra, one of the harbours conceded to Germany on the Persian Gulf. In March, 1903, a German-controlled syndicate was authorised to complete its construction. The construction rights were accompanied by oil rights extending up to a 20 kilometre strip on either side of the track.[8]

As the expanding influence of Germany threatened to dominate the region, Anglo–Russian rivalry was replaced by a mutually shared anxiety. For the British, the threat of German land power in dangerous proximity to the Near and Middle Eastern focal points of the Empire's lifeline, strengthened the belief that the advance of Germany in the region could best be met by coming to terms with Russia. Similarly, the growing German involvement in the region, together with the humiliating defeat in 1905 in the Russo–Japanese War, forced the Russian government to review its position in the Near East. As a result a convention was signed in August 1907 between Britain and Russia which created an entirely new situation in the region. The differences of the two powers over Asian territories were settled, the way was paved for their later alliance in the First World War.[9]

The Ottoman Empire on the other hand, realising that it could no longer count upon Great Britain to protect it against Russia, drew closer to Germany. In turn Russia, following the 1907 Convention, opened negotiations with Britain in the hope of securing some arrangement for passage from the Black Sea. The Anglo–Russian Entente remained in force until November 1917[10] when the Bolshevik Revolution overthrew the old regime and abolished its treaties and obligations. My interest in the early history of modern Turkey is that of an international historian. This work explores the territory where research in international relations and post-war diplomacy interfaces with enquiries on the Turkish war of independence and the nationalist movement. I have integrated these hitherto largely separate projects into a single discussion. In doing so, the intention is to analyse not only the international rivalry in the region but also to shed a new light on some of the fundamental assumptions of early Turkish history. Particular emphasis is placed on the restraints and dilemmas confronting the Turkish nationalists in their search for an independent Turkish homeland under the pressure of Western imperialism and the pull of Russian Bolshevism.

The primary aim of this book is to provide a documentary account of the struggle for power and influence between two major powers, Britain and Soviet Russia, in the region. The work concentrates on the years 1918 to 1923, a short but vital period in the history of the lands from Thrace to the Caucasus. These five years represent a crucial phase. Not only were the overall forms of international relations in the area defined during these years, but also the foundations of the regional political structures and the orientations of various local social and political interests were established.

These years were characterised by war and revolution, Allied intervention and civil war in Russia, foreign invasion and the Greco–Turkish war in Anatolia. Grand empires were crushed to death and new states were created. Territories changed hands and governments toppled. Borders shifted following the fighting in the battle field and manouvres at the diplomatic table. It took years to fix them permanently.

This work examines on the one hand the development of the events in the Caucasus and Turkey within the context of the role played by Soviet Russia and Britain. On the other, I also analyse the specific response of each power to the events in Turkey and the Caucasus and the implications of these developments within each country's broader foreign policy concerns. In this way the investigation attempts to show how domestic developments in Turkey and the Caucasus affected the actual conduct of foreign affairs in London and Moscow, and vice versa.

These questions constitute the overall boundaries of the study. The actual narration of the events is organised on three distinct but related planes of action: the rivalry between Britain and Soviet Russia over establishing their respective spheres of influence; the rapprochement between the Bolsheviks and the Turkish nationalists; and British policy in the post-war Turkish settlement.

While treating the three different manifestations of the post-war settlement within one text it is natural that the scope of the investigation is kept limited and focused as much as possible around central questions. It is therefore deliberately confined to that part of the British–Soviet competition which took place in the context of the long-term settlement of the Turkish question. Similarly, only those parts of the Turco–Soviet relations that relate to British intervention in the region are included in the narration.

Focus throughout is on the political aspects of British–Soviet rivalry in the region. The tide of military confrontation, both in the Russian Civil War and in the Greek–Turkish War, is not treated in detail since it was regarded as secondary to the political battle and the making of foreign policies. The same point must be made with regard to the domestic policies of the British and Soviet governments. While their domestic conditions and their inner party struggles are all a part of the general environment of specific developments, they are discussed only in so far as they impinge directly on their rivalry and respective involvement in Turkish and the Caucasian affairs.

The text is organised chronologically into six chapters. The events of the first year after the Bolshevik Revolution are examined in Chapter 1. The main focus of this part is on the decision of the new regime in Petrograd to end hostilities on the fronts, which provided a real chance for the Turkish army to exercise its influence in the Caucasus without serious opposition.

Chapter 2 narrates the formal ending of the war and aims to illustrate how the Turkish armistice of October 1918 brought Turkey and the Caucasus under the control of the British. It continues to show the critical shift in the regional balance which led to the British decision to withdraw troops from the Caucasus and to concentrate forces in Turkey.

Chapter 3 attempts to capture the accelerated pace of events following the British decision to pull the troops out of the Caucasus. During this period the region of the Caucasus witnessed increasing Bolshevik activity, leaving the British alarmed. Turco–Soviet co-operation constitutes the central axis of the chapter, which traces the course of the major events both in Turkey and the Caucasus from March 1919 to March 1921. This section ends with the consolidation of the Soviet power in the Caucasus.

Chapter 4 deals with the Anglo–Soviet Trade Agreement and the Turco–Soviet Treaty of Friendship, both signed on the same day (16 March 1921). In this section the negotiations leading to the agreements and the immediate impact of both agreements are discussed.

While the first phase of the British–Soviet rivalry had concluded with the Sovietisation of the Caucasus by mid-1921, many of the problems persisted afterwards and undoubtedly influenced the course of events in Turkey. Chapter 5 thus covers a complicated and eventful period largely dominated by the military campaigns in Anatolia.

Finally, in Chapter 6, 'Settlement', the main aspects of the disparity between the designs of each side and the terms of the final settlement are discussed. This chapter accounts for the re-emergence of the familiar issues between Russia and Britain and gives a description of the territorial issues resolved at Lausanne.

1

The Red Star and the Green Crescent November 1917–September 1918

> The Soviet Government must straight away offer to all the belligerent peoples (i.e. simultaneously both to their governments and to the worker and peasant masses) to conclude an immediate general peace on democratic terms, and also to conclude an immediate armistice (even if only for three months).
>
> The main condition for a democratic peace is the renunciation of annexations (seizures) – not in the incorrect sense that all powers get back what they have lost, but in the only correct sense that every nationality without any exception, both in Europe and in the colonies, shall obtain its freedom and the possibility to decide for itself whether it is to form a separate state or whether it is to enter into the composition of some other state.
>
> In offering the peace terms, the Soviet Government must itself immediately take steps towards their fulfilment, i.e., it must publish and repudiate the secret treaties by which we have been bound up to the present time, those which were concluded by the tsar and which give Russian capitalists the promise of the pillaging of Turkey, Austria, etc. Then we must immediately satisfy the demands of the Ukrainians and the Finns, ensure them, as well as all other non-Russian nationalities in Russia, full freedom, including freedom of secession, applying the same to all Armenia, undertaking to evacuate that country as well as the Turkish lands occupied by us ... [Lenin, October 1917][1]

The Bolshevik Revolution and the Eastern Front

When the Bolshevik Revolution started in Russia the First World War was already approaching its fourth year. Although the Russian armies had controlled the regions of Transcaucasia and part of eastern Anatolia since 1916, on the European borders of Russia control was in the hands of the Central Powers. The armies of Germany and Austro–Hungary had invaded Russian Poland and had reached the borders of the Ukraine.

It was no surprise that the urgent task which the new regime in Petrograd set for itself was that of peace. After four years of war the Russian people were weary. The soldiers particularly wanted to return to their homes. In Lenin's words, 'the soldiers are tired out, the soldiers are barefoot, the soldiers are starving, the soldiers do not want to fight.'[2] In this context it is understandable that in the days preceding the November Revolution the Bolsheviks had promised peace. If peace was not forthcoming, if thousands of men could not return home, it was obvious that Lenin's fate would be no different from that of Alexander Kerensky (the last prime minister of the Provisional Government). It was the pro–war commitments of Kerensky, as described by Taylor, which 'opened the door to the Bolsheviks by attempting to revive the war'.[3] The Bolsheviks had to make peace if they were to survive, and this they did.

The Decree on Peace[4] was the first act of foreign policy of the new government. Through it, Lenin proposed to all those peoples and their governments who were engaged in war the immediate opening of negotiations for a democratic and just peace without annexations and indemnities. He also stated that 'the secret treaties must be published' and 'the clauses dealing with annexations and indemnities must be annulled.'[5] Lenin further proposed an immediate armistice for three months in particular to 'the class conscious workers of the three most advanced nations of the world' – England, France, and Germany.[6]

Britain and its allies considered the Bolsheviks' call for peace, with its direct implications on the fronts, as an adverse blow to the Allied war effort. This first act of the new regime presented a radically different attitude from its predecessor, Kerensky's Provisional Government. Russia opted out of the war front as a major power. For Britain this meant the end of Anglo–Russian co-operation in the East, with the immediate danger of leaving south Russia and the Caucasus wide open to enemy influence. In the case of the Bolsheviks' making a peace with the Central Powers, it would be highly optimistic to expect decisive results against Germany and its allies in near future.[7]

With the new regime in Russia there emerged a fear in the Allied bloc that the disintegration of the Russian army would leave Germany and its allies virtually unopposed at the Eastern front. The Germans had 76 divisions on their eastern front in March 1917, and the Allies did not want these to be moved to the West. The transfer of as many as 2 million German soldiers from the Eastern to the Western front raised the distinct possibility of a German victory.[8] It seemed vital, therefore, to keep the Germans and their allies occupied in the East.

Recognition of this threat started a hasty search for a means of filling the vacuum in the East[9] and for this purpose the Romanian Army based on the north-eastern town of Jassy seemed to many to be a possible candidate. In the aftermath of the Bolshevik Revolution the British War Cabinet discussed the matter.[10] Although no decision was reached, the cabinet decided to seek the advice of Colonel House, the head of the American Mission and special representative of President Wilson in Europe. House was in London for preliminary discussions before an early December meeting of the Supreme War Council in Paris. He believed that the most the Allies could do would be to advise the Romanians to co-operate with whichever forces loyal to the Allied cause were nearest them. This policy was adopted by the War Cabinet on 22 November 1917 as an immediate response.[11]

At that time the only Russian forces loyal to the Allies were the troops of A. M. Kaledin, the Cossack general at Novocherkassk on the Don. General Kaledin was the commander of the 8th Army on the south-western front and the head (*ataman*) of the local Cossack congress, the Voisko Cossack Circle.[12] In September 1917 Kaledin had broken with the Kerensky regime after siding with General Kornilov, leader of an unsuccessful attempt to seize power.[13] The day after the Bolsheviks took power he refused to recognise the new government and proclaimed an independent regime at Novocherkassk.[14]

Soon further information arrived from the region. On 29 November 1917, General Knox, British military attaché in Petrograd, reported the formation of a 'South Eastern Union' formed by representatives of all the important Cossack armies in south Russia.[15] On 29 October–2 November, less than a week before the armed uprising in Petrograd, a conference of the Don, Kuban, Terek, Astrakhan, Kalmyk, Ural and Daghestan Cossack units and the representatives of the nomadic tribes of the North Caucasian steppe had been held in Vladikavkaz to establish the 'South Eastern Union' (or League). Its aim was described as the provision of mutual support for the purpose of maintaining order and legality within the Union until the All-Russian Constituent Assembly met.[16]

This Cossack initiative was significant, for it represented the first important attempt to unify at least a portion of Russian lands outside Bolshevik control. The initiative created a hope that if the Union could be strengthened and supported materially, it would be very useful to the Allies in the region. With the help of the Cossacks, the Georgians and Armenians could be reinforced against the Turks at the Caucasian front. This was considered by the Allies to be an important opportunity to continue the war in case of a separate peace.[17] These Christian forces of the South, if organised, could occupy the Donets Basin and thus keep the German and Turkish forces from getting coal, iron, or oil from Russia or grain from Siberia.[18]

The question of supporting loyal Russian groups in the Don region and also in Transcaucasia was raised at the Inter-Allied Conference convened on 30 November in Paris. The meeting decided to send a combined Anglo–French military mission from Romania to Kaledin's headquarters.[19] However, the British foreign secretary, Arthur James Balfour, was not pleased with this decision. According to Balfour it was undesirable to dispatch a mission to Kaledin until reliable reports were received from the region. On 9 December, in a memorandum, he expressed his concerns, arguing that hasty steps might 'drive Russia into the hands of Germany'.[20]

Despite Balfour's reservations, Lloyd George and the rest of the War Cabinet decided to give financial guarantees to Kaledin, regardless of the expense. Lloyd George describes in his memoirs why he supported this decision: 'This attitude, if successful, would have averted the worst disasters of Russian defection.'[21] This was believed to be one way to keep Russia in the war, even after the Bolsheviks signed a separate peace with Germany. Although at this early stage there was no satisfactory evidence about the strength of those Russian elements loyal to the Allied cause, to have optimistic hopes regarding the influence of such forces was considered at least a policy rather than not having one.[22] Page, the American ambassador to London, reports that although the British leaders 'confessed to a lack of knowledge themselves of the situation' in south Russia, they considered the available anti-Bolshevik forces 'the only chance' to continue the war.[23]

While the British were actively occupied over how to keep Russia in the war against the Central Powers, the Bolshevik government was busy trying to establish some basis to declare an immediate armistice on the fronts. As a first step, on the night of 20 November the Council of People's Commissars[24] sent a wireless message to General Dukhonin,[25] the commander-in-chief at General Staff Field Headquarters in Mogilev, and directed him with a proposal to his German military counterpart of a three months' armistice.[26] The next day, Trotsky, then people's commissar for foreign affairs,

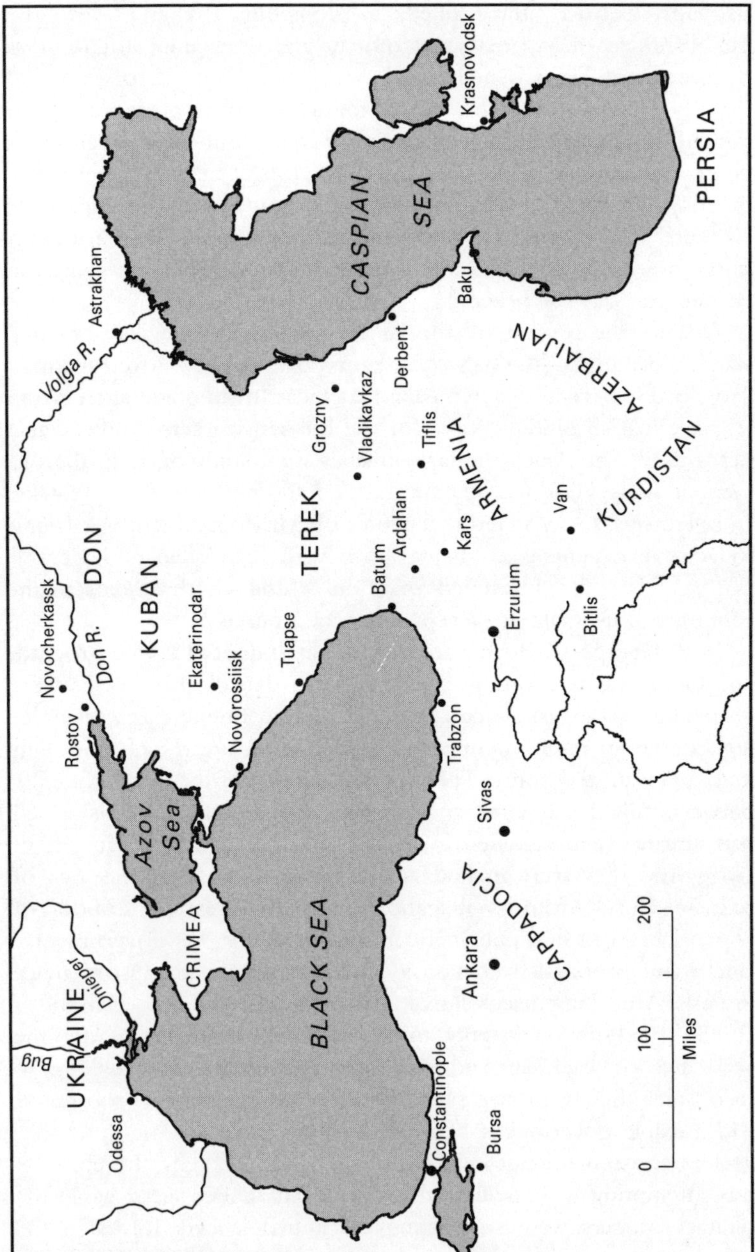

Map 1 South Russia

formally notified the Allied missions in Petrograd of the establishment of the new government and drew their attention to the Decree of Peace issued on 8 November by Lenin. Trotsky added that the decree should be considered as a formal proposal for an immediate armistice on all fronts and for the immediate opening of peace negotiations.[27] On the same day Trotsky made a speech at a meeting of the Central Executive Committee of the Soviet of Workers' and Peasants' Deputies. In this speech he singled out Britain as the power that would initiate a most hostile attitude regarding the Bolsheviks' call for peace.[28]

Despite the protest of the Allies that the Bolsheviks' initiative was a violation of Russia's obligations to the Allied governments,[29] Trotsky disregarded this by saying that those 'treaties and agreements ... have lost all binding force for the Russian workers, soldiers, and peasants'.[30] The Allied military attachés were summoned to the old General Staff Headquarters and handed another note from Trotsky. In it Trotsky clearly stated that the Bolshevik government was trying to achieve a general armistice, both with the Allied and Central Powers. However, he stated that the refusal of the Allies might force the Bolsheviks to accept a separate peace.[31]

Both George W. Buchanan, the British ambassador to Petrograd, and General Knox were aware of the fact that the Bolsheviks were desperate for an armistice and would not continue the war under any condition. Large numbers of soldiers had already left the front lines and returned home. They realised that it was the incompatibility between Allied war aims and Russia's own domestic needs which had already been resolved in the Bolsheviks' favour in November 1917. And they were inclined to come to terms with the idea of releasing Russia from its obligations under the Treaty of London of September 1914 – a course Buchanan urged on his government.[32] Buchanan's proposals were considered at the Inter–Allied Conference in Paris. After long discussions it was finally decided that each power should send its own reply to its ambassador at Petrograd, the substance of each answer to be that the Allies were willing to reconsider their war aims in conjunction with Russia as soon as it had a stable government with whom they could act.[33]

The apparent dichotomy in British policy whereby the embassy was attempting to conciliate the Bolsheviks in Petrograd while the military officers were supporting the anti-Bolshevik forces in the provinces[34] caused increased concern in the War Cabinet. To clarify the British position, on 21 December the War Cabinet adopted a memorandum on Balfour's proposal. The next day Milner, a member of the War Cabinet, and Cecil, the minister of blockade, left for Paris with this memorandum to submit to the French premier, Clemenceau, and Pichon, the French foreign minister. According to the memorandum, the Allies were to make it clear to the Bolsheviks that they had no desire to take any part in Russian internal politics.

The memorandum also stated that continued regular contacts with the Ukraine, the Cossacks, Finland, Siberia, and the Caucasus would be necessary on both the grounds of moral obligation and the strategic importance of these regions. Balfour's memorandum particularly urged that the contacts with the anti-Bolshevik movements and their armies 'should be done as quietly as possible, so as to avoid the imputation … that we are preparing to make war on the Bolsheviks.'[35] This memorandum was accepted by Clemenceau the next day and became the Anglo–French Convention.[36] In this convention two central elements of the Allied policy were put forward: on the one hand to prevent Germany and its allies from obtaining rich material sources (especially the oil of Azerbaijan and the wheat of the Ukraine)[37] and on the other to stop the escalation of the Turkish influence in the region. It was emphasised that a Pan-Turkic and Islamic expansion from Turkey into the heart of the Caucasus might be an even greater danger to the peace of the Western world.[38]

This concern about 'the development of a Turanian[39] movement that will extend from Constantinople to China'[40] was related to another, increasing fear: The Ottoman government, encouraged by the Bolsheviks' Decree of Peace, was expecting to recover most of its eastern Anatolian and Transcaucasian lands occupied by the Russian armies since the summer of 1916. The German–Soviet negotiations provided the Ottoman government with an opportunity to put forward further claims in the Caucasus where a considerable Muslim population had already been stirred up by nationalist sentiments.[41] In this way the possibility of an imminent Turkish expansion in the region presented 'a new and a very real danger' to the British position in the East.[42]

The Anglo–French Convention of 23 December 1917 specifically addressed the issue of Turkish expansion. This convention divided the area into spheres of influence. The Cossack territories, the territory of the Caucasus and Kurdistan were assigned to Britain; Ukraine and Bessarabia were allocated as the French zone. It was clearly stated that the British were to concentrate their efforts mainly south of the Caucasus mountains, and against the Turks. The immediate aim was to keep these territories and their resources from the Germans and the Turks.[43]

In order to establish firm control over those regions of south Russia and the Caucasus, the British simultaneously pursued various actions. The War Cabinet approved large sums of money to the anti-Bolshevik Russian forces to continue resistance on the Ukrainian, Romanian and on the Don fronts, as well as in the Caucasus.[44] At the same time, Allied agents and intelligence officers were sent in to advise and support the governments of Transcaucasia and the anti-Bolshevik armies which would continue the war.[45]

In this way, soon after the Bolshevik revolution, the Caucasus emerged as 'the most vital [area] from the point of view of British interests'.[46] The British Empire had long considered the region strategically vital to its power in Persia, Afghanistan and, most important of all, India. When the Bolsheviks proposed an armistice the expectation was that the region would fall into chaos. This in turn would make the area vulnerable to the influence of both German and Turkish armies, who would surely attempt to grab the golden opportunity to put the region under their control. The possibility of losing the Caucasus to the Central Powers had serious consequences for Britain's regional as well as global interests. The Caucasus could provide the Central Powers with countless material resources, including the rich oil reserves of Baku, the coal mines of Tkibuli and Tkvarcheli in Georgia, the manganese mines of Chiatura (in Georgia), copper in Armenia, and iron of Azerbaijan. This would counteract the Allied blockade against Germany and its allies and, combined with the expected transfer of German troops from the Eastern to the Western front, delay the anticipated victory by the Allies.

With the fear of losing the strategic positions and the valuable natural resources of the Caucasus to the Central Powers, the British government encouraged the Christian nations of the region to erect a firm barrier. At the same time they provided material assistance to the anti-Bolshevik Russian armies.[47] Even if the Bolsheviks had reached a separate peace agreement with the Central Powers which would open the doors for the enemy advance in the area, the Allied-supported bloc was hoped to prevent the fall of the key routes to the German and Turkish powers.

Direct British involvement in the region was initiated by the strategic concerns arising from the deterioration of the Caucasian front. It was a preventive action aimed to protect the south Russian and the Caucasian regions against the threat posed by the German and Turkish armies. The Bolsheviks' persistent call for peace was similarly considered in the context of its likely contribution to a

German and Turkish victory in the region. From the British point of view, as far as the south Russian and Caucasian territories were concerned, Bolshevism had merely cleared the way for the Germans and the Turks.

The Bolshevik government in Moscow, on the other hand, considered the involvement of the British in south Russia and the Caucasus as a direct military act against the sovereignty of the Soviet state,[48] although the Bolsheviks did not have the control of the region at that time. British support for the anti-Bolshevik forces was regarded as an openly hostile act. Britain was accused of initiating a well-organised and ideologically motivated war which aimed 'to destroy the power of the Soviets'. It was described as a 'class war' initiated by 'the British imperialists' who 'were embittered and frightened by the revolution'.[49] In this way, the relationship between the new Soviet regime and the Western powers was at first moulded as much by the exigencies of war as it was by ideological differences.

Turkish Advance in the Caucasus

By the end of 1917, it appeared that British fears were well-founded. All the indications from the region pointed to an immediate occupation of Transcaucasia by the Ottoman armies in a fresh attempt to realise their long-standing dream of unifying the Turkic-speaking Muslim peoples of the Caucasus, Transcaspia and Central Asia under the Turkish banner. Indeed, the high expectations of the Turks had been manifested even before the Bolshevik Revolution in Russia. Immediately after the March Revolution, Talat Pasha,[50] the Ottoman premier, had stated that the abolition of the Tsarist regime opened the doors to the realisation of Turkey's eastern empire.[51] According to the Sublime Porte, the revolution had already proved that the long-dreaded enemy, the Russian Empire and its army, had disintegrated.

In Constantinople there had been expectation of an immediate peace on the Russian front. This was reflected in the reports of the Ottoman press as well as in the government circles. However, by mid-1917 it had become apparent that the peace would not come with Kerensky's Provisional Government. From June onwards the attention of the Turkish press had shifted to the Bolsheviks as the party that would bring the long-awaited peace.[52] Following the Bolshevik Revolution, the sense of anticipation in the Sublime Porte reached a new climax as the Russian armies were expected to withdraw from the Eastern front. Some newspapers were escalating the euphoria, with leading articles calling for immediate recovery of the lands in the eastern Anatolia and Transcaucasia.[53]

The decision of the new regime in Petrograd to end hostilities on the fronts strengthened Turkish expectations and provided a real chance for the Turkish army to exercise its influence in the region without any serious opposition. General Ludendorff, the deputy chief of the German General Staff, describes in his memoirs how the Turks seized the opportunity to fill the resulting political and military vacuum in the Caucasus as being 'the predestined moment' for the realisation of all their ambitious schemes of Pan-Turanian expansion.[54] During the peace negotiations between the Central Powers and Soviet Russia, which started in December 1917, Turkish expectations regarding the Caucasus significantly increased. Soon after the negotiations started the Turkish representatives put forward a demand for the Ardahan, Kars, and Batum *vilayets*,[55] which had been ceded to Russia under the Berlin Treaty of 1878. It was considered a very convenient moment by the Sublime Porte to recover those strategic eastern *vilayets* and to unite the Anatolian Turks with the Muslim population of the Caucasus.[56]

In line with the contemporary press and popular sentiments, the Ottoman government had grasped the importance of the opportunity even before the Brest–Litovsk negotiations, and initial steps had been taken to acquire a significant footing in the Caucasus. Some 'even saw the mirage of a new empire in the Caucasus'.[57] Enver Pasha,[58] Ottoman war minister and the most-celebrated champion of the Pan-Turanian expansion, had sent two divisions from the Galician and Moldavian fronts to the Turkish–Russian border in the east during the last year of the war. The intention was to create a sound military base in the Caucasus in case of an Allied-occupation of western Turkey.[59]

The Turkish delegation at Brest adopted a dual policy: namely to secure an agreement with the Soviet government which would provide them a legal framework for their claims in the region, and at the same time to wait for the Bolsheviks' acceptance of the harsh German demands to put forward further claims regarding the regions of Transcaucasia.

The treaty of Brest–Litovsk, signed on 3 March 1918, provided the much appreciated and diplomatically tailored legal framework for the Turkish armies to advance in the Caucasus. Despite the demoralising defeat of the Turkish armies on the Syrian, Mesopotamian and European fronts, the Turkish delegation at Brest secured significant gains. Three key *vilayets*, Ardahan, Kars and Batum, would be separated from Russian territory. This was done under the pretext of the right of nations to dispose of themselves. Following the signature of the treaty, the US minister in Sweden quoted the Swedish press commenting that the Turks would recover what was lost to Russia during the nineteenth century, and most importantly of all:

> Kars, of strategic importance as barricading fort in mountains, but Batum is place of most value to Turkey as pipes run there from great petroleum centre Baku on Caspian Sea. The tremendous petroleum trade at Baku will come under Turkish control.[60]

In the treaty of Brest–Litovsk (article IV),[61] Russia promised to 'ensure the immediate evacuation of the *vilayets* of the eastern Anatolia and their lawful return to Turkey'.[62] A supplementary agreement, titled the Turkish–Russian Treaty, appended to the treaty of Brest–Litovsk regulated the execution of this provision.[63] According to this treaty, Russia was forbidden to concentrate more than one division, even for drill purposes, on the borders of the three *vilayets* or in the Caucasus without previous notice to the Central Powers, until the conclusion of a general peace. Turkey, on the other hand, was allowed to keep its army on war footing. As had already been emphasised in the Swedish press, the importance of this retrocession of the three *vilayets* lay not so much in the return to Turkey of the great fortress of Kars but in the fact that Batum was the key port for the rich Baku petroleum.[64]

When the Bolsheviks at the Brest–Litovsk Conference ceded Batum, Kars and Ardahan to Turkey this caused a deep anxiety in the Caucasus. The region itself, as elsewhere in Russia, was on the brink of chaos after Kerensky's downfall in Petrograd. The whole area, situated between the Black and Caspian Seas, is divided by the chain of the Caucasian Mountains into two distinct parts: north Caucasus and Transcaucasia. In the north Caucasus the representatives of the Terek Cossacks, following the decision of the South-Eastern Union, refused to recognise the Bolshevik regime in Petrograd. The Mountain Peoples of the north Caucasus and Daghestan followed suit, and, as a result of a series of joint meetings, the central authorities of the Terek Cossacks and the Mountain Peoples set up a 'Provisional Terek–Daghestan Government'[65] in early December.[66] The Provisional Terek–Daghestan government assumed authority in the entire area of the north Caucasus, the capital of the new republic being Vladikavkaz, the Russian city at the edge of the mountains on the end of the Georgian Military Road.[67]

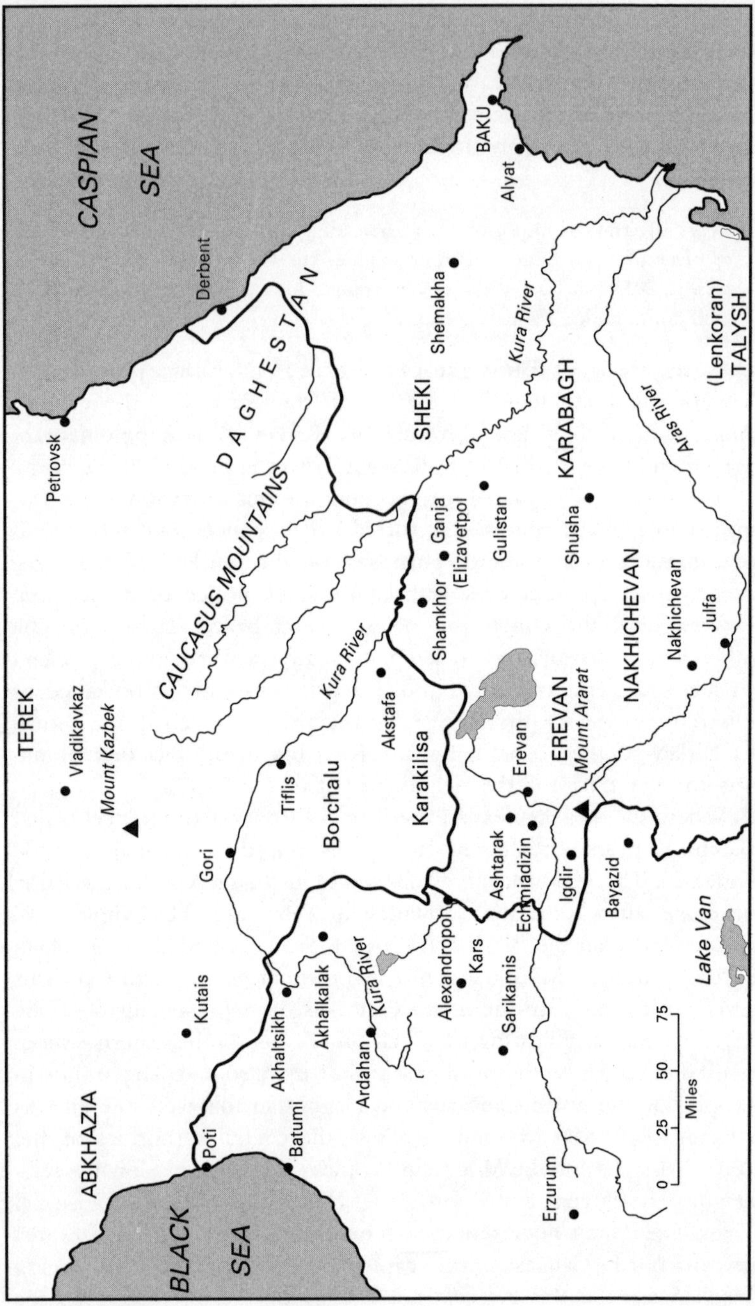

Map 2 The Caucasus

South of the Caucasus mountains lay Transcaucasia, extending over an area of 74,000 sq. miles. Georgians, Armenians and Azerbaijanis, together with a number of minority groups, inhabited this area. In November when the Bolsheviks seized power in Russia, a meeting of the representatives of Georgia, Azerbaijan and Armenia took place in Tiflis and established an interim government for the region under the title of the 'Transcaucasian Commissariat'. The Transcaucasian Commissariat, or Zakavkom, included three Georgians, three Azerbaijanis, three Armenians, and two Russians. Gegechkori, a Georgian Menshevik, was elected president of the Commissariat and commissar of external affairs and labour.[68]

The new Transcaucasian government declared its intention to retain power until the Constituent Assembly could convene. The delegates of the Transcaucasian Commissariat began to function independently immediately after the dissolution of the Russian Constituent Assembly by the Bolsheviks on 19 January 1918.[69] A representative assembly, the Seim, was formed with the participation of the representatives of Georgia, Armenia and Azerbaijan. In April 1918 the Seim proclaimed Transcaucasia an independent Democratic Federative Republic and a cabinet was formed with executive powers.[70] The Seim did not recognise the outcome of the Brest–Litovsk conference; in fact it had attempted to enter separate negotiations with the Sublime Porte in February while the discussions were still going on at Brest–Litovsk.[71] Increasing activities of the Turks in the region accounted for this decision. The Ottoman troops were concentrating in large numbers and Russia, one of the most powerful of the Allies in the region, was already negotiating a humiliating peace with the Central Powers. Under these circumstances it might have appeared that the most realistic way to survive was to come to terms with the Turks. Two days before the conclusion of the Brest–Litovsk treaty, the Seim chose a delegation to negotiate with the Turks in the name of the Transcaucasian Assembly.[72]

Transcaucasian and Turkish delegations met at Trabzon, a south-eastern Black Sea port, to discuss peace terms at a conference which started on 14 March.[73] When ıe supplementary Turkish–Russian agreement to the treaty of Bre c–Litovsk, which had already ceded Batum, Kars and Ardahan to Turkey, was put forward by the Turkish delegation as a precondition for the talks, the negotiations entered a deadlock.[74] The Transcaucasian government mobilised the army in order to prevent what they perceived as an imminent Turkish invasion of the three *vilayets*. After stubborn fighting, the Turkish forces penetrated the eastern part of Transcaucasia where they conveniently joined hands with the Azerbaijani Muslims who were not enthusiastic supporters of the Seim after all.[75]

While the Turkish armies were slowly advancing from the south following the deadlock in the Trabzon negotiations, 300 miles away in Baku there were major clashes between the local Muslims and the Bolsheviks. Baku was the only town in Transcaucasia under strong Bolshevik control. Following the March Revolution in Petrograd a 'dual power'(*dvoevlastie*) emerged in Baku as well as in the other big cities of Transcaucasia. On 4 March, following the directives from the Provisional Government in Petrograd, an 'Executive Committee of Public Organisations' (Ispolnitel'nyi Komitet Obshchestvennykh Organizatsii) was established in Baku to act as the local organ of the new central government. Two days later the Baku Soviet of Workers' Deputies had its first meeting with the participation of 52 delegates, representing more than half of the Baku proletariat, and elected Stepan Shaumian as chairman *in absentia*.[76] Shaumian returned two days later from exile in the Volga town Saratov and assumed the leadership of the Baku Soviet. The Bolsheviks in Baku did not follow the example of revolution in Petrograd but chose instead to gain power locally by long-term political change in local government by 'peaceful transition'.[77]

Baku was a town with a population of 2,500 at the beginning of the nineteenth century. By the end of the century it had emerged as an industrial and commercial centre with a thriving population of 200,000. The rush to Baku was driven by the discovery of rich oil resources enabling Tsarist Russia to become an exporter of petroleum. In 1893 Russia became the largest oil-producing country in the world. Railways soon followed and the Batum–Baku railway construction opened the way for a big harbour. By 1918 Baku was at the centre of every important transit route between Russia, Iran and Central Asia via the Caspian Sea.

The population of Baku consisted of three major nationalities – Russians, Armenians, and Muslim Azeris. Muslims accounted for more than half of the labour force in the oil industry. The better jobs, requiring skills and training, were held by Russians and Armenians. These ethnic communities lived in distinct neighbourhoods: wealthy merchants, businessmen and professionals (among whom few were Azeris) lived in the centre of the city. Azeris inhabited hastily built shanty towns in the industrial suburbs and the more distant oil field districts.[78] Each ethnic group had its own political organisation. The Armenians centred around the Dashnaktsutiun, a socialist party with a strong emphasis upon Armenian national unity;[79] the majority of the Azerbaijanis developed some loyalty to the secular nationalist Musavat Party;[80] and the Russians were close to the SRP and the Mensheviks. There had been a rising sympathy towards the Bolsheviks since the March Revolution. The real issue in Baku was the danger of an ethnic war between the Armenian and the Muslim communities.

When the Caucasian front disintegrated completely following the Bolshevik Decree of Peace, there remained virtually no army on which the Baku Soviet could rely. As a result, an uneasy co-operation emerged between the Bolsheviks and the Armenian Dashnaktsutiun. The Bolsheviks had no choice but to rely on the volunteer units of the Armenians. This new development increased ethnic and religious hostilities and further alienated the Muslim community in Baku.

Within this unstable atmosphere the Muslims took up arms following the fall of the Kerensky's Provisional Government. The core of the Muslim forces was the Tatar cavalry regiment of the Touzemnaia (Native) Division, unofficially known as the Dikaia (Wild) Division,[81] transferred from Petrograd. Starting from the beginning of 1918, sporadic fights broke out in the region between Muslims and Christian Russians and Armenians. The situation became extremely tense in Baku when the Armenian military units became the chief political ally of the Bolsheviks.

Conflict flared in late March. On 24 March a shipload of Muslim soldiers of the Wild Division arrived in Baku and resisted orders of the Baku Soviet to disarm. The Armenians first declared their neutrality as a Muslim rebellion started against the authority of the local Soviet. But the clash immediately developed into racial and religious warfare when the Armenian units joined forces against the Muslims. Within a day the Bolsheviks, supported by the Armenian units, fought a very violent war against the Baku Muslims led by the Musavat. The fighting lasted three days and Musavat was defeated. There were heavy casualties, mostly Azerbaijani Muslims. According to Shaumian's figures more than 3,000 Muslim volunteers were killed.[82]

This episode is described in Soviet historiography as 'the victory of revolutionary forces' against the 'counter-revolutionary forces' of the Musavatists, the Dashnaks, the SRs and the Mensheviks. The Armenian support in the Baku Bolsheviks' fight against the Muslims was hastily swept aside in order to justify the incident in terms of a 'class struggle' between 'Transcaucasia's working class, and above all, its vanguard the Baku proletariat' and the 'Azerbaijanian bourgeois-gentry circles'.[83]

The March events in Baku became the first tragic incident in which the Bolsheviks sought the support of one ethnic power against another to strengthen their position in the region. Following the end of the battle, the Bolsheviks hurriedly started to consolidate their gains. With the Muslim power tragically cut and the Armenians weakened by the battle, there was no one to oppose the full monopoly of power by the Bolsheviks. They absorbed part of the Armenian troops into the Caucasian Red Army, disbanded the rest, and closed all opposition parties and papers.[84] The Baku Council of People's Commissars (Sovnarkom) was set up at a meeting of Baku Soviet on 25 April.[85] It declared itself as the first plenipotentiary Soviet government body in Transcaucasia.[86] Thus Baku's 'October' arrived six months after Petrograd's. For the Azerbaijani Muslims March 1918 represents a decisive turning-point. Following their heavy losses they rejected the new authority in Baku and severed all their links with the Bolsheviks, putting their hopes in the advancing Ottoman armies as their potential liberators from their non-Muslim rulers.

While the Baku Bolsheviks were busy in consolidating their gains after the March events, the Ottoman armies continued their advance from the other side of Transcaucasia towards the Caspian shores. By mid-April they had already occupied Batum and Kars. In the face of this rapid advance the government of the recently declared Transcaucasian Republic accepted the call of the Turks for a new conference in Batum, then under the Ottoman occupation.[87]

The Batum conference started in early May. At the conference the Ottoman delegation presented a draft treaty which demanded from the Tiflis *guberniia*, the Akhalkalak and Akhaltsikh *uezds* and from the Erevan *guberniia*, the entire Surmeli *uezd* along with part of the Alexandropol and Etchmiadzin *uezds* through which the Kars–Julfa railroad passed.[88] In addition to these new territorial claims, the draft of the Sublime Porte included substantial privileges in trade and navigation, frontier traffic, full Ottoman transit rights through Transcaucasia, and a sharp reduction of the armed forces of the Transcaucasian government in Tiflis.[89]

The Batum Conference marked the apex of Turkish activist visions. During the negotiations Vehib Bey of the Ottoman delegation demonstrated the Turkish point of view as:

> You see that destiny draws Turkey from the West to the East. We left the Balkans, we are also leaving Africa, but we must extend toward the East. Our blood, our religion, our language is there. And this has an irresistible

> magnetism. Our brothers are in Baku, Daghestan, Turkestan, and Azerbaijan. We must have a road toward those areas.[90]

There seemed nothing to prevent the Turks from placing the whole region under their control. The Azerbaijani Muslims were opposed to any resistance as they were only too happy to see a Turkish take-over. The Armenians, although fearful of such a possibility, were exhausted and disorganised and therefore not able to participate in any kind of opposition. Finding themselves in a desperate situation and fearing complete isolation, the Georgians approached the German General von Lossow with a request for protection.[91]

The Germans were interested in forestalling the advance of the Turkish armies as they were seeking control of petroleum and other natural sources in the area for their own benefit. The Georgian demand provided an opportunity for the Germans to set up a strong basis in the region to secure 'the raw materials of the Caucasus independently of Turkey'.[92] Colonel Kress von Kressenstein, representative of the Imperial Chancellor, together with two German officers and a detachment, arrived in Tiflis and established close contacts with the members of the Georgian Diet.[93]

As a result of this guarantee provided by Germany the Georgian Diet assembled in Tiflis and declared independence of Georgia on 26 May.[94] With this declaration of independence, after only a month's existence, the Transcaucasian Republic came to an end. Two days later, the independent republics of Armenia and Azerbaijan were declared, with Erevan as the Armenian capital, and Gandzha (Elisavetpol) the Azerbaijani.[95]

The first government of independent Georgia was headed by a Menshevik, A. Chkhenkeli, and clearly identified with a pro-German policy. Khan Khoiski formed the first Azerbaijani cabinet in which most of the posts were controlled by the Musavat. Having almost no military force of its own, independent Azerbaijan was, however, totally dependent upon the Ottoman army commanders. Political authority in Armenia was firmly in the hands of one party, the Dashnaktsutiun; but unlike the Georgians and the Azerbaijanis, the Armenians had no one to whom to turn for assistance.

On 4 June 1918, three republics signed three separate agreements with the Ottoman Government at Batum, thus ending the Batum Conference. They accepted the harsh terms of the Turks in the hope of winning time by yielding space. By the Treaty of Brest–Litovsk the Ottoman Empire had acquired nearly 10,000 square miles and 600,000 inhabitants of Transcaucasia. As a result of the Batum

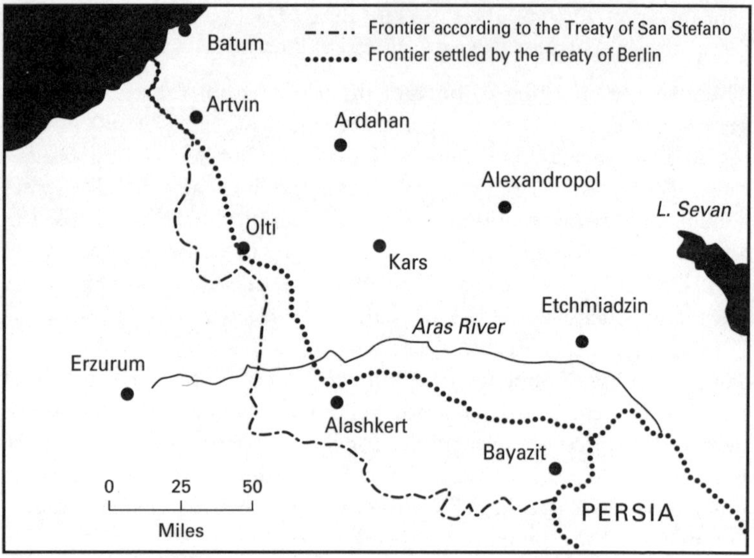

Map 3 Russo–Turkish Frontiers

agreements an additional 8,000 square miles and 650,000 people were added to the Turkish sphere, delivering fresh blood to the ailing Empire.[96] In this way, by June 1918, the Turks recovered in the Caucasus not just the 1914 Russo–Turkish frontier but also the 1878 (Berlin Treaty) frontier.[97]

Article IV of the Agreement with Azerbaijan, which accepted that 'the Ottoman Government will provide military support to the Azerbaijani Government if this is seen necessary by the latter for domestic stability and national security,' signified the special relationship between the Turks and the Azerbaijanis.[98] Following this agreement a Turkish army, together with Azerbaijani and other Muslim volunteers from the region, began to move towards the Caspian against the Russian Bolsheviks and the Armenians who controlled Baku since the tragic 'March days'.[99] The Turkish army, called the 'Army of Islam', was under the command of Enver's younger brother, Nuri Pasha, who was known for his extreme pan-Turanian ideas. The first task of Nuri's army was to secure a route from Kars through Julfa to Baku. Once secured, it would then be possible to control the regions of north Persia as well.[100]

In north Persia the British forces had been actively involved in operations since the beginning of the year, with the goal of reaching the Caspian front from the south to establish contacts with the pro-

Allied elements in Transcaucasia. These operations were directed by Major-General Lionel C. Dunsterville, the hero of Kipling's novel *Stalky and Co.*[101] Dunsterville had been appointed chief of the British mission to the Caucasus and also British representative at Tiflis in January 1918.[102] His sphere of work was also to cover all Russian and Turkish territories south of the main chain of the Caucasus. The primary military objective of Dunsterville's mission was defined as 'the maintenance of an effective force on the Caucasus front so as to protect the occupied portions of Turkish Armenia and to prevent the realisation of Pan-Turanian designs'.[103] To achieve this objective Dunsterville was to cross the 650 miles between Baghdad and Enzeli and from there to embark his men for Baku and Tiflis.[104]

The preparations for Dunsterville's advance in the Caucasus were undertaken by a small number of British military intelligence officers who by then had set up a committee in Tiflis called the 'Caucasus Military Agency' under the control of Colonel G. D. Pike. This was a small wartime mission attached to the Tsarist Caucasian headquarters at Tiflis. After the Revolution the members of the mission had stayed on, trying to persuade the Russians to continue resisting the Turks.

Major G. M. Goldsmith, who was to assist the operations of Dunsterville as 'I', had arrived in Baku on 13 February 1918 with two Ford cars and three men, and soon after proceeded to Tiflis to work with the Caucasus Military Agency. Goldsmith had set up relations with the local groups in the region, including the Bolsheviks, and established an efficient communication link with the British forces in north Persia.[105]

Starting from early June, Turkish troops were coming closer to the line Dunsterville was trying to hold.[106] With the extremely favourable terms of the Batum conference the Turkish army had gained a greater foothold for further operations in the region.

In this situation the Bolsheviks in Baku and the British in north Persia found themselves under a common threat stemming from the advancing Turkish troops. If the Turks captured Baku and the surrounding oil fields, it seemed obvious that the expectations of the British and ironically those of the Bolsheviks would come to a disappointing end. Once the Turks were established in Baku it would be very difficult to prevent a Turkish advance further eastward into Persia and Transcaspia. One had only to cross the Caspian at its narrowest point to reach Krasnovodsk and the Turkoman steppe, from which could be reached the vast expanse of Turkestan.

It was, therefore, not surprising for the British and the Bolsheviks to share a common goal in May 1918: to keep the Turks away from Baku and surrounding oil fields. As the Turkish forces under the leadership of Nuri Pasha were marching against Baku, neither of the two parties could afford to be selective about their allies. Neither the Bolsheviks nor the British had sufficient ground forces in the region to stop the Turkish armies. The small number of British troops could only act as a nucleus for some form of reorganisation of the local forces against the Turkish advance and unless they achieved a kind of understanding with the Baku Bolsheviks this too seemed difficult to realise.[107] Indeed, it was one of the first successes of Goldsmith's mission in the region to get permission from Shaumian in late February to let Dunsterville, together with 40 officers and 50 other ranks with 4 motor cars, to pass through Baku and proceed to Tiflis.[108]

Dunsterville does not mention the Caucasus Military Agency in his memoirs.[109] Neither does he make any reference to Shaumian's permission. But on 22 May he asked General Marshall, general officer commanding-in-chief of the Mesopotamian Expeditionary Force in Baghdad, for permission to collect all available troops and armoured cars to help the defence of Baku against the Turkish forces. On 24 May General Marshall repeated this demand in a telegram to the War Office stating that Dunsterville, with all available force, would be ready to leave for Baku within a week.[110]

Despite the general recognition that there was very little Britain could do by itself, once asked by their military representative in the region the British authorities in London found it very difficult to give consent to any such co-operation with the Bolsheviks. On 27 May the War Office sent a telegram to the Mesopotamian H. Q. saying that 'in view of the completely changed situation in the Caucasus, General Dunsterville was not to go there'.[111] To this Dunsterville replied that 'he did not wish Baku to give up hope and allow the oil to fall into enemy hands.' He repeated his earlier demand to give every assistance possible to the defence of Baku. This was repeated in a telegram to the CIGS sent by General Marshall on 4 June.[112]

The CIGS replied General Marshall's telegram on 6 June saying that 'without further reference to London and definitive instructions from there' no step was to be taken.[113] The issue was discussed in the War Cabinet and the Eastern Committee.[114] The conclusion of these discussions was the same with the CIGS's instructions to General Marshall: it was not desirable to send the British troops to Baku.[115]

The reasons for this refusal were probably threefold. In the first place, the Eastern Committee feared that once the troops had gone there they would inevitably be overwhelmed by the vastly more numerous Turkish forces. Second, Lord Curzon, the committee's chairman, insisted to his colleagues that the foundation of their policy had been to build a barrier against the Turks in northern Persia and if Dunsterville transferred any of his forces to Baku it was feared that the Persian barrier would be weakened.[116] Finally Lloyd George provided the ultimate and perhaps the most significant reason why the British could and would not co-operate with the Bolsheviks. In a War Cabinet meeting in June he stated that 'it would be better for us for the Turks to hold Baku' rather than the 'Russian Bear', since he believed that 'it was not probable they [the Turks] would ever be dangerous to our [British] interests in the East, whilst, on the other hand, Russia, if in the future she became regenerated, might be so.'[117]

Interestingly enough Moscow's reaction to the issue of inviting the British to help the defence of Baku against the Turks was exactly the same as London's. The Bolshevik government refused to allow Dunsterville to pass unimpeded to Baku. Instead of handing Baku to the British imperialists it was preferred to let the Turks capture it. Like London, Moscow believed that Turkish rule in the city would not be permanent. The Baku Bolsheviks were urged to oppose categorically the idea of inviting British troops for the defence of the city.[118]

The official opposition both from London and Moscow, however, did not stop their local representatives from considering such co-operation and establishing some contacts. Goldsmith recounts in his report that the Baku Bolsheviks 'actually assisted' the Caucasus Military Agency 'in preventing food transports' by the Germans. In Vladikavkaz too the Bolsheviks co-operated with the Caucasus Military Agency to check the German and Turkish advance in the north Caucasus. Goldsmith provides Russian and English copies of a detailed action plan prepared by Colonel Pike alongside the Vladikavkaz Bolsheviks. According to this the Georgian Military Road, together with the railway station in Vladikavkaz, were controlled by a combined team of the British and Bolsheviks 'against the German and Turkish agents ...'[119]

In addition to Goldsmith's account Major Ronald MacDonell provides some extra confirmation of this in his personal account of the events. MacDonell had served as British vice-consul in Baku for seven years. Before that he had lived there for a number of years.

He was on familiar terms with most of the leading figures in the region. In December 1917 MacDonell was given the rank of major and seconded to the British military mission in Tiflis for 'special duties'. According to MacDonell the leader of the Baku Bolsheviks, Shaumian, had a number of meetings with him in May and, although reluctantly, considered inviting the British forces to Baku to help the city's defences.[120]

There were other examples of co-operation between the local Bolsheviks and the British. As a result of the negotiations in Enzeli, Dunsterville arranged with the local Bolshevik committee to sell Ford cars to the Baku Soviet in exchange for fuel. Fifty thousand pounds of gasoline was bought in return for ten Fords by the end of July.[121] The barter of ten Ford cars with petrol was cited by Colonel Rawlinson as well. Rawlinson was from 7 July onwards under the orders of Dunsterville.[122]

Throughout this time the Turkish army was rapidly advancing towards Baku and the city was surrounded by mid-June. Regardless of the Bolsheviks' refusal, the Armenian Dashnaks, who actually constituted the majority of the Baku Red Army, sent a delegation to the British General Dunsterville and asked for help in defence of Baku against the Turks.[123]

First, Colonel Lazar Bicherakhov, the anti-Bolshevik Russian leader with whose 1,800-strong Cossack-force Dunsterville had been co-operating, went and offered his troops to the Baku Soviet. Both Dunsterville and Rawlinson indicate that this offer was clearly initiated and implemented by Dunsterville as a substitute after London refused his proposal to take the British troops to Baku to organise the defence of the city together with the Bolsheviks.[124] In his letter to the Baku Council of People's Commissars Bicherakhov wrote: 'I am not a Bolshevik or Menshevik. I love my country, know how to fight a bit and am coming to help the Baku Soviet defend the city of Baku from the Turkish invasion.'[125] Despite his record of anti-Bolshevik activities, the Baku Bolsheviks did not oppose this offer and Bicherakhov was welcomed in Baku.[126]

Bicherakhov and his regiment, accompanied by a few British officers, with two armoured trains, artillery, and armoured cars, reached Alyat, a port some 35 miles south of Baku, early in July and immediately started to combat with the Turks.[127] He was appointed commanding officer of one of the chief units of the Baku Red Army.[128] On 9 July the Baku *Izvestiia* was carrying a confident message from Bicherakhov saying that the chances of the Turkish army were small in the face of the young Red Army.[129] However, this optimism was short-lived. Bicherakhov's troops were outnumbered and they could not hold the front against the Turkish infantry and artillery for long. After holding back the Turks for about a week, he withdrew his troops from the front and retreated to Derbent and Petrovsk in Daghestan.[130]

The Turkish forces on the other hand were pursuing their march with even greater determination. By the end of June new divisions from the Western front had arrived in the Caucasus, where three Ottoman armies were waiting for the instructions of Enver Pasha. The Third Army (composed of the 3rd, 5th, 36th, and 37th Caucasian divisions) was charged with maintaining order in all territories acquired by the treaties of Brest–Litovsk and Batum. The newly organised Ninth Army (made up of the 9th, 10th, and 11th Caucasian divisions) was bestowed upon Yakub Sevki Pasha, whose temporary headquarters were in Alexandropol. Together, the Ninth Army and the Sixth Army (the latter located in North Persia), constituted the Army Group of the East under the supreme command of Enver's uncle, Halil Pasha. The ambitious task of liberating Baku and expelling the British from Persia and Baghdad rested with these troops. Altogether these were some 55,000 to 60,000 soldiers and several thousand irregulars under Ottoman control.[131]

Already Mürsel Pasha's 5th Caucasian Division was in transit to Gandzha, where it was to form the nucleus of Nuri's 'Army of Islam' and bolster Ali Agha Shikhlinskii's Azerbaijani forces. The Germans, having reached a preliminary understanding with the Ottomans, withdrew their units to a line of villages north of the Kamenka River – thus allowing Mürsel's men to pass from Alexandropol and Karakilisa into the Elisavetpol *guberniia*. Kress von Kressenstein now counselled the Georgian government to let the Turks trespass in the southernmost parts of the Tiflis *guberniia*.[132]

On 16 July 1918 the Dashnaks in the Baku Soviet, this time together with the SRs, put another motion to extend an official invitation to the British. The proposal was narrowly defeated but it had received so much support that Shaumian decided to ask for instructions from the Bolshevik central authorities. Moscow repeated its previous decision, and ordered Shaumian to combat the Baku Soviet's 'unpardonable disposition' to appeal for British aid. Were Baku to fall, it would be better that the Ottoman Turks – not the experienced English colonialists – become the temporary heirs to the invaluable city and its resources.[133]

During the last couple of weeks of July 1918, the possibility of a Turkish invasion of Baku drew ever closer. The arrival of new divisions from the Western front reinforced the capability of the Turkish forces. It also became clear that the Germans would not try to prevent the Turks from pursuing their advance on Baku. The Bolshevik leaders of Russia could offer little help. According to Soviet historians the Bolshevik government in Moscow, with the directives of 'the Great Leader', was considering sending a significant number of Red Army troops to defend Baku.[134] But in the end only a cavalry force of about 170 men came from Astrakhan and a detachment of 780 infantry from Central Asia.[135]

On 25 July the Turkish armies reached Alyat, and on the same day at the meeting of the Baku Soviet (despite the opposition of Shaumian) a final motion to seek British military aid was voted and accepted by a small margin (259 to 236).[136] Shaumian declared that he regarded the decision as 'a disgraceful betrayal'[137] and, as the representative of the central Soviet power in Baku, he declined to burden the responsibility. Six days later the Bolsheviks withdrew from the Baku Soviet.[138] On the following day this was replaced by the Centro-Caspian Directorate – a coalition dominated by SRs and composed of the Russians and the Armenians.[139] As its first act the Centro-Caspian Directorate officially invited Dunsterville to protect Baku. Dunsterville's long-awaited moment had finally arrived.[140]

On the basis of this publicly announced appeal from Baku Dunsterville apparently persuaded London that with the support of a small British force Baku could successfully resist the Turkish armies. He was given permission to send a total of two British battalions, with supporting artillery and armoured cars.[141] On 4 August 1918 the first British battalion arrived in Baku. As its first directive London instructed Dunsterville not to hesitate 'to dispose of any remaining Bolshevik influence he might find at Baku.'[142]

During the next couple of weeks more British soldiers landed in Baku. Among them were 70 British troops who arrived with an empty oil-tank steamer on 17 August. Colonel Rawlinson was in charge of this detachment. As soon as he arrived in the British headquarters he was given the title of 'Controller General of the Ordinance' by the Transcaucasian Directorate. Rawlinson immediately started a widespread campaign to collect all available weaponry, ammunition and explosives for the use of the city defences.[143]

It is difficult to give the exact number of the British troops who arrived in Baku in August. Despite the variation in the quoted number of men in various accounts,[144] it appears that British forces on the ground had been around 1,000 men even by the most exaggerated account. This small force was totally inadequate even for the purpose of compelling the Turks to raise the siege of Baku.[145] Dunsterville attempted to strengthen the defence of the city by commanding the units of the Dashnaks, SRs and Mensheviks. He soon came to the conclusion that it was not realistic to protect Baku unless extensive military support was provided. Yet none seemed forthcoming.[146]

In late August a glimpse of hope emerged more than 1,000 miles away to save Baku from the invasion of the Turkish armies. On 27 August 1918 the German–Soviet talks ended with an additional treaty to the Brest–Litovsk. In this treaty the articles related to the Caucasus proclaimed that Soviet Russia would not oppose German recognition of Georgia, and that Russia would deliver to Germany either a quarter of the petroleum tapped in Baku or a certain monthly minimum to be determined later. In return, Germany would refuse to assist the military operations of a third power in areas beyond the borders of Georgia and, most importantly of all, Germany would strive to prohibit a third power from entering the Baku region.[147] On paper this supplementary treaty was a serious setback to the advancing Ottoman armies. However, the number of German troops in the Caucasus were not enough to stop the Turks militarily. Furthermore the existing German troops and officers were just called back due to 'the great efforts made against us [the Germans] on the French front'.[148]

The Turks were only 35 miles away from the city of Baku. In order to complete a de facto situation the 5th, 15th, and 36th divisions of the Caucasian Turkish army were immediately ordered to strike rapidly. In the face of this strong Turkish attack the German military leaders did not attempt to fulfil those relevant articles of the Brest–Litovsk treaty. They did not deny the Turkish troops' entrance to Baku and even they recommended that a small number of German units should participate in the campaign in order to safeguard some share for themselves.[149]

By the end of August 1918 Dunsterville had definitely decided that 'the further defence of Baku is a waste of time and life.'[150] He addressed the Baku authorities on 1 September and repeated the same point: 'No power on earth can save Baku from the Turks. To continue the defence means only to defer the evil moment and cause further needless loss of life.'[151] London was informed about the situation in Baku on 28 August. On 31 August the War Office telegraphed in reply expressing entire concurrence for a full withdrawal. It was also suggested that Dunsterville should be instructed to destroy the oil-plant at Baku before withdrawal.[152]

When on 12 September 'definite and reliable' information arrived (which according to Dunsterville came from two Arab deserters from the Ottoman army and according to Rawlinson from an Armenian) about the Turkish plans for the final attack, Dunsterville made the preparations for evacuation.[153] On the night of 14 September, with Baku under heavy attack, Dunsterville loaded his men on ships and left the city to its fate. Rawlinson, on the other hand, managed to load most of the ammunition on to a little ship and left the city with large quantities of explosives. Despite the request of the War Office regarding the Baku oil-plant no damage had been done before evacuation, beyond putting the wireless station at Baku out of action.[154] The same night the Turks broke through Baku's final defence parameter. The next day it was declared that the combined Turkish–Azerbaijani army captured Baku on behalf of the Azerbaijani government.[155] A new government was established at Baku, under the leadership of Khan-Khoiski who had just arrived from Gandzha.[156]

Dunsterville, Rawlinson and MacDonell blamed the local forces – especially the Armenians – for the fall of Baku. The Armenians according to their accounts were 'undisciplined', 'disorganised', 'lacked fighting spirit', and 'were unable to understand anything about warfare'.[157] Some Armenian participants of the events, however, challenged this view by claiming that Dunsterville had initially promised them to bring a much larger force to Baku. Sergei Melik-Yolchian, a member of the Centro-Caspian Directorate, states that Dunsterville had promised that a British army of 5,000 was ready at Enzeli to embark for Baku, yet the total number of the British troops and technical personnel in Baku was hardly more than a thousand.[158]

One of the British officers of the adventure, Captain Reginald Teague-Jones[159] of Military Intelligence, indicates in his diary that 'in practice the venture was doomed to failure because of two main factors (among many others): the force was too small for the task assigned to it, and it arrived much too late.'[160]

The occupation of Baku by Turkish troops marked the end of the first phase of the turbulent events in the Caucasus. The fall of Baku was a significant loss both for the British and the Bolsheviks. The Turkish control of this strategic post meant a certain setback for the British position in the region by opening the wealthy oil resources of Baku to the exploitation of the Central Powers within threatening proximity of the British empire in Asia. For the Bolsheviks this episode signified the end of their only power base in the whole region of the Caucasus while at the same time increasing the potential dangers stemmed from an anti-Bolshevik Muslim campaign towards Transcaspia and Central Asia.

In the face of the advancing Turkish armies both Moscow and London officially refused to co-operate. This was despite the fact that, as Ullman states in the first volume of his *Anglo–Soviet Relations*, 'there was an assumption' on the British side that there did 'exist a community of interest between the Allies and the Soviet regime'.[161]

As illustrated by the events of Transcaucasia from May to September 1918 there indeed existed such an assumption, not only on the British side but also among the Bolsheviks. In the face of the immediate danger of a Turkish take-over of Baku and surrounding oil fields both the representatives of the British and those of the central Soviet government in the region found themselves to be on the same side. Both wanted to keep the Turks out of Baku. Despite the existence of major differences both sides came close to assuming that they had a common interest at that vital point. Co-operation based upon reciprocity could develop between local Bolsheviks and the British in the region.

The policy-makers of each side, however, did not share the same view. They were too busy to read the long-term implications of the other government's policy. Therefore, the assumption that both Britain and Soviet Russia had a common interest did not exist at the governmental level. The British involvement in the Caucasus was seen by Moscow as nothing but a clear manifestation of the British imperialists' anti-Soviet campaign with the ultimate aim of destruction of the socialist regime in Russia. The British, on the other hand, still considered the Caucasian events as part of the old 'Great Game' along nineteenth-century lines:[162] an imperial concept of the power struggle for the Asian provinces between Britain and the 'Russian Bear'.

This mutual fear and suspicion, as it existed in Anglo–Russian relations up to the Bolshevik Revolution of 1917, was also dominant in the minds of the policy makers in London and Moscow after November 1917. This resulted in both sides instructing their representatives not to enter into any kind of co-ordination to protect Baku from the Turkish invasion – thus leaving the men on the spot with a lack of direction. In this atmosphere of confusion and hesitancy each side sought to counter-balance the Turkish threat in its own pragmatic way. While the local Bolsheviks were trying to ally one national power (Christian Armenians) against the Muslim Azerbaijanis, the British hoped that an anti-Turkish, anti-German bloc (which included the Russians loyal to the Allied cause and the Christian nations of the region) could be formed.

The actions of the local Bolsheviks and those of the British in the Caucasus did not follow well-planned decisions but consisted rather of day-to-day tactical moves. These were pragmatic operations aimed to take advantage of the opportunities presented by the local circumstances. The local conditions themselves were changing fast, urging quick responses sometimes conflicting with each other. As a result, out of this confusion the control of the region easily fell to the Turks.

2

The End of the War
October 1918–March 1919

> A single mandatory for the Turkish Empire and the Trans-Caucasus would be the most economical solution. No intelligent scheme for development of railroads for Trans-Caucasia and Armenia can be worked out without extension into Anatolia ... [H. W. V. Temperley][1]

> What we had to do for the moment was to prevent disorder in the Caucasus, and give a chance to the autonomous states that were struggling into existence ... we must keep troops there and keep open the Batum–Baku railway ... [Curzon, 12 December 1918][2]

The Mudros Armistice

Less than two months after the Turkish invasion of Baku the First World War officially ended and the rule of the whole area passed from the Ottoman and German armies into the hands of the victorious British and their allies.[3] With all of its war-time rivals defeated and the Bolsheviks fighting for their own existence in the Civil War, Britain was rewarded with a unique opportunity of re-shaping the future of the lands stretching from the Caspian in the East to the European hinterland of the Black Sea Straits in the west.

For the Turks the realisation of the defeat had come earlier in September during a visit of the grand vizier, Talat Pasha, to Germany. Talat Pasha, on his return from Germany, had witnessed the collapse of the Bulgarian Army which had lost the battle of Dubropolje to the multi-national *Armée de l'Orient* under General Franchet d'Esperey. The Bulgarian government had officially notified Talat Pasha when his train passed through Sofia that Bulgaria would seek a separate peace with the Allies. On 30 September Bulgaria left the war, and the Ottoman government was left in frustrated isolation with no direct means of communication with its war allies, the Central Powers in Europe.[4]

Faced with the coming defeat the Sublime Porte realised that a new government not identified with the war-time policies would be better placed to negotiate with the Allies. A new cabinet was formed at Constantinople on 17 October in which there were a number of members who were outspoken opponents of Turkey's alliance with Germany.[5] The new grand vizier, old chief-of-staff Field Marshall Ahmet Izzet Pasha,[6] was one of the most respected, highest-ranking, military leaders of the Empire. He had recently led the Turkish delegation at the Brest–Litovsk peace discussions with the Soviet government. Most importantly he was not a member of the Committee of Union and Progress [the CUP] – the ruling party since 1913 which was blamed for both the war and the humiliating defeat.[7]

As its first task the new cabinet delegated Hüseyin Rauf,[8] minister of marine, to negotiate the armistice with the British commander-in-chief in the Mediterranean, Vice-Admiral Sir Somerset Arthur Gough-Calthorpe at port Mudros on Limnos island in the Aegean Sea.[9] Hüseyin Rauf had been at the Brest talks as the representative of the Ottoman navy and he was also well known for his opposition to the German influence in the Ottoman army and navy.[10]

During the negotiations at Mudros the Turkish side tried to secure certain concessions.[11] In the original draft the Ottomans were required to withdraw their forces from all those eastern regions which they had hastily occupied during the final throes of the war.[12] The eastern lands in question were north Persia and all Transcaucasia which, of course, included Baku and the valuable *vilayets* of Batum, Kars and Ardahan. For these three *vilayets* the Ottoman delegates took a forceful, uncompromising, stand. Calthorpe, however, agreed to compromise on other parts of eastern Anatolia but not the strategic Transcaucasian provinces.[13] Batum especially was regarded as the key point for effective control in the region for it provided a major strategic terminus in the Caucasus with its rail and pipeline connections to the oil fields around Baku.

By the terms of the armistice, the defeated Ottoman army had therefore to withdraw from north-west Persia, Baku, Batum and the rest of Transcaucasia. According to Article XV all Transcaucasian railways were also put under the control of the Allies.[14] Nor did the Mudros negotiations yield any compromise to the Turks over the Straits. Securing Allied access to the Black Sea was considered of supreme importance since command of the Black Sea would make it possible to secure positions in the Caucasus, the Caspian and the south Russian regions.[15] It was, therefore, not a coincidence that the very first article of the Armistice with Turkey opened the Straits and thus 're-established our [Allied] liaison with South Russia'.[16]

To implement Article XV, on the day immediately following the conclusion of Armistice discussions with Turkey, the War Office ordered the British Command in Mesopotamia to occupy Baku and its surrounding oil fields. Acting Major General William Montgomery Thomson, the commander of the British expeditionary force in north Persia, ordered the last Ottoman contingents out of Baku on 14 November 1918. Thomson entered Baku from north Persia on 17 November, exactly 64 days after Dunsterville's troops had left the city to the Turks.[17] Meanwhile, other Allied troops from the Salonica front (including three British divisions) arrived in Constantinople and settled on both sides of the Bosphorus, Sea of Marmara and Dardanelles. A British naval squadron was also dispatched to the Ottoman capital to secure the direct control of the sea routes and the Straits. The commander of the Allied forces was Vice-Admiral Somerset Gough-Calthorpe, who became the British High Commissioner. He lived on HMS Iron Duke moored off the Princes' Islands.[18]

The control of the whole area, from the Straits to the Caspian, was assigned to the recently established Army of the Black Sea. The three British divisions, which arrived in Constantinople from the Salonica front, became the backbone of this army which was put under the command of General George Francis Milne.[19] From the headquarters of the Army of the Black Sea in Constantinople Major General G. T. Forestier-Walker's 27th Division was dispatched to safeguard the vital centres in the western provinces of Transcaucasia.[20]

In the Ottoman capital a British-led Allied administration was established by December 1918. Allied Control Commissions supervised demobilisation and disarmament of the Ottoman armies, and regulated the local police and gendarmerie. The city was divided into spheres of influence among the Allied powers – with the major consideration of sharing the financial burden associated with the control of such a big metropolis with more than 900,000 inhabitants.[21] Britain had jurisdiction in the Galata and Pera suburbs, on the north shore of the Golden Horn where foreign residential housing, the consulates, and fashionable shops were located; the French had the control of Istanbul, the old Turkish city where the historical mosques, Turkish government buildings and traditional bazaars were located; the Italians had authority in Üsküdar, the residential district on the Asian shore of the Bosphorus; and a small

Greek contingent was in control in the Phanar.[22] In addition to the capital, British forces occupied several coastal towns along the Straits and the Dardanelles, Samsun (a major harbour city on the Black Sea coast), Antep in the south-east, and some other strategic points, as well as the whole length of the German-built Anatolian railway.[23] Turkey was at the mercy of the Allied and Associated powers'.[24]

In this way the Turkish armistice put two neighbouring regions – the Turkish lands in Thrace and Anatolia and the Caucasus – under British military control, carried out by the Army of the Black Sea. The geographic proximity of the Caucasus to Anatolia made it both physically possible and strategically important for the British to establish control in both areas. As a result of this reinforced unity, Britain had the opportunity of increasing the intensity of its influence as well as its geographical control.

The end of the war effectively destroyed the power centres in the region by excluding Russia, Germany and Turkey from the political scene for the foreseeable future.[25] The immediate problem was the political vacuum that had emerged in the region. The most likely solution to fill the gap was for Britain to impose 'order' as the victorious force. The question was, however, how to maintain 'that order' once it was enforced.

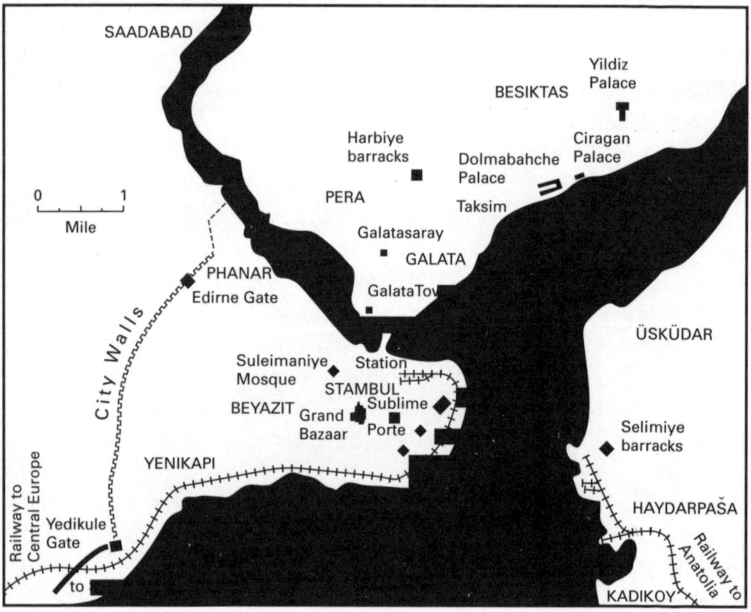

Map 4 Constantinople

Turkish Affairs

The following months showed that the initial superiority based on victory was not enough for effective control. There was no unity of purpose and much less commitment in the victors' camp. Progress towards a long-term settlement was weakened from the beginning by the increasing rivalry among the Entente powers and the lack of unity in the British government. There was tension between the British high commissioner and the high commissioners of France and Italy. This was partly because the dominant role of the British was regarded with suspicion by France and Italy. In Constantinople, the cream of the ancient city, the most prestigious areas of Pera and Galata were under British control and their allies were not happy with this arrangement. A series of crises over Allied control in the region emerged soon after its establishment in Constantinople, and exacerbated the delicate complications of the post-war Turkish problem.[26] The British government considered that it was natural and fair for them to deserve a special status with a leading role in the Turkish lands. 'We have,' said the British prime minister, 'taken by far the larger part of the burden of the war against Turkey in the Dardanelles and in Gallipoli, in Egypt, in Mesopotamia and in Palestine ...'[27] Now it was time to reap the rewards.

However the British were divided over their role in Turkey. There existed a wide range of often conflicting opinions among the British politicians, military leaders and diplomatic representatives in the region. The main confusion came from the fact that the end of the war also ended the consensus held in Lloyd George's coalition government over the Turkish question. The armistice triggered a period of disunity which undermined the stability of the government.[28]

The British government's policy on Turkey had been originally outlined by Lloyd George and his foreign secretary, Lord Balfour. Balfour had engineered a number of secret agreements among the Allied powers during the war which aimed at the complete dissolution of the Ottoman Empire. The Constantinople Agreement of March/April 1915 between Britain, Russia and France had acknowledged Russia's demand to annex Constantinople and the Straits in the event of an Entente victory. The Treaty of London of April 1915 between the Entente powers and Italy pledged Italy a portion of the Mediterranean region bordering the province of Adalia [modern Antalya]. The Sykes–Picot Agreement of May 1916, offered the total partition of the Ottoman Empire among Britain, France and Russia.

And the St Jean de Maurienne Agreement of April 1917, clarified the Italian demands, indicating Italian take-over of the Aydin province together with Smyrna [modern Izmir].[29]

Balfour, in strict accord with these war-time commitments, was keen to pursue a policy of stripping the Turks of their empire and of banishing them from Europe. They should, he believed, be ejected from Constantinople to a small central Anatolian state that would be reserved for them,[30] and Constantinople should be placed under international control or a mandatory power. Lloyd George was in agreement with Balfour in principle. He also advocated a policy which would not only punish Turkey for her role in the war but also bring a decisive conclusion to the long-standing 'Eastern Question' by putting the key provinces of the Ottoman empire under Allied control.[31] However, Winston Churchill, minister of munitions after July 1917 and secretary of state for war after January 1919, had a different opinion of the Turkish issue. For Churchill the problem had wider implications and therefore needed a comprehensive strategy. It was not only a matter of enforcing a small Anatolian Turkish state as a solution to a long-standing problem; there was the equally important issue of the newly emerging Bolshevik power in Russia. With this in mind Churchill was quick to re-formulate his strategic priorities in favour of a different policy. Although he was, like Balfour and Lloyd George, in favour of non-Turkish areas being detached from the Turkish state, he believed that there were a

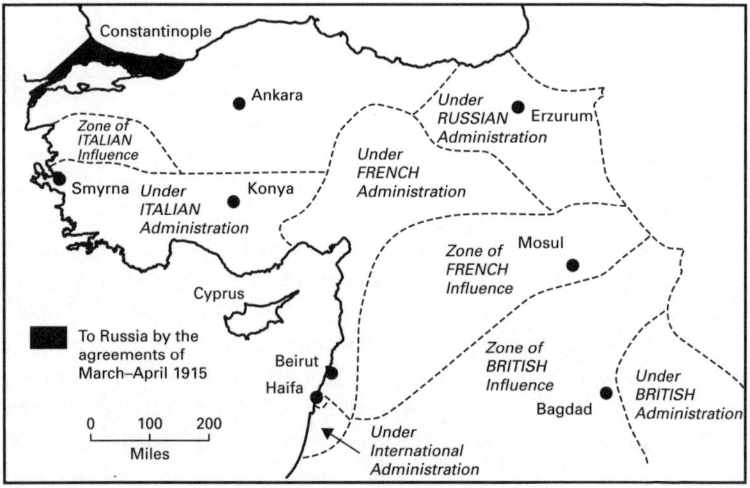

Map 5 The Division of Turkey according to Secret Agreements

number of advantages in allowing the Turks to keep Constantinople under their banner.

Churchill believed that there were practical, historical and strategic reasons to justify a softer approach to the Turks. Firstly, it would be much easier to maintain control of Anatolia and the eastern Mediterranean with Constantinople as the capital of a friendly Turkish state. Secondly, because of the strong historical and cultural links between Turkey and the Muslims in India, he was concerned that the expulsion of the Turks from their capital might produce a backlash in India. Finally Churchill believed that an Allied occupation of Constantinople may drive the Turks into a strategic alliance with the Bolsheviks against the British.

Ernest S. Montagu, the secretary of state for India, was another strong critic of the government's policy and shared most of Churchill's concerns. He feared that a harsh approach to Turkey might strain British rule in India. Montagu indicated his desire for a relatively fair settlement to the Turkish question as early as October 1918.[32] In his letter to Maurice Hankey,[33] secretary to the Cabinet, Montagu wanted him to communicate to Lloyd George the hope that the Turks should be allowed a relatively fair peace settlement similar to the one with Bulgaria.[34]

In the British Cabinet one politician was distinguished for his clarity of vision with respect to Turkish affairs. Lord Curzon, then the acting foreign secretary, and foreign secretary following Balfour's resignation in October 1919, had a vision that encompassed not just the Turkish lands but the region as a whole. He argued for 'a chain of friendly states stretching from the confines of Europe to the frontier of the Indian Empire …'.[35] The future of the Turkish state was the central link in this visionary chain. Curzon proposed a detailed programme for his project. According to the 'Curzon Plan' Arab lands would be separated from the Turkish state and would be subject to the principle of 'self-determination'. However, he supported Balfour and Lloyd George over the issue of the Turkish capital – the Turks, he thought, should be expelled from Constantinople due to its strategic location. The city was viewed as a springboard from which to control the outcome of events in south Russia, the Caucasus and Transcaspia. For the remaining territories in Anatolia he disagreed with Lloyd George and Balfour, and was anxious to assure territorial integrity and no partition.[36]

Despite the existence of different opinions within the cabinet and the rivalry between the Allied powers, Lloyd George was quite optimistic. When in the early days of 1919 statesmen representing the victorious powers in the First World War gathered in Paris to draft peace terms that were to be imposed upon their defeated enemies, Lloyd George strongly believed that the Turkish question would be solved in accordance with his own considerations.[37] 'The partition scheme' is perhaps the best term to describe the British prime minister's approach to the Turkish settlement. According to this scheme the power vacuum that had previously been filled by the Ottoman Empire in Thrace, Constantinople and the eastern Mediterranean should now be filled by a friendly power. Britain itself could not undertake this commitment, as it was witnessing a powerful campaign for demobilisation[38] and the War Office believed that there were not enough troops or funds available to pacify Turkey. The British military authorities had thus requested that the British Government should look for a proxy to police Turkey.[39]

Who would have volunteered for such a task? Greece emerged as almost a natural candidate. When the Greek delegation appeared at the Peace Conference they seemed to Lloyd George the most appropriate choice. In his view the Greeks would be the people of the future in the eastern Mediterranean. He expected that as 'they were prolific and full of energy', they would be best suited to the task. A greater Greece, to Lloyd George, would be an invaluable advantage to the British Empire. If Greece, a nation of five or six million, he thought, could hold the territories which have been assigned to them, in 50 years they would be a strong nation of 20 million.[40] This reasoning was, indeed, very much in line with the claims of the Greek delegation headed by Prime Minister Venizelos[41] in Paris. The Greek case was passionately presented to the Peace Conference by Venizelos on 3 and 4 February 1919.[42] Venizelos's claims were quite detailed.[43] He admitted that the Turkish population of western Thrace far surpassed that of Greeks and Bulgarians combined.[44] However, he claimed the Turks would prefer Greek rule to Bulgarian. To support this claim Venizelos presented a long memorandum signed by the prominent Turkish leaders of Bulgarian Thrace asking for the union of the entire province with Greece on the ground that Bulgarian rule was unbearable.[45]

Venizelos put forward additional claims in western Asia Minor concerning the islands off the coast, part of the province of Bursa, all of Aydin province with the exception of one borough, Denizli, and above all the *vilayet* of Smyrna.[46] According to Venizelos these territories had a population of 1,188,359 Greeks and 1,042,050 Muslim Turks, and the area was described as Aegean and therefore Greek, rather than a part of the Asiatic hinterland in terms of climate, geography and culture.[47] On the issue of Constantinople the Greek delegation was deliberately cautious and preferred to be silent. Such a policy, they believed, would strengthen their claims to the rest of Thrace.[48] The Greek delegation advocated the internationalisation of the city under an international mandate. However, they also believed that 'owing to the strength of the Greek community Constantinople would become a small Greek state under international protection and the fulfilment of the innermost national desires would follow naturally.'[49] As an initial tactic the Greeks did not demand the city, yet in the long run they were anticipating its gradual take over.[50] The major problem with Greek claims originated from the fact that they were against the interests of another member of the alliance, Italy, which claimed the right to annex the islands of Rhodes and Dodecanese, and a share in the mainland particularly in the *vilayet* of Smyrna. These claims had been endorsed by the Allies in both the 1915 Treaty of London and the St Jean de Maurienne Agreement of 1917.[51]

In early 1919, under the proposal of Lloyd George, the Council of Ten[52] in Paris decided to establish a committee of experts to examine the Greek claims and put forward recommendations. The committee began its investigation in February. Britain and France accepted the Greek position for most part, while Italy bitterly opposed the scheme and the United States adopted a more reserved stand.[53]

Mainly because of Italian opposition, no final decision on Greek claims was reached among the Allies. The Italians were isolated and they became increasingly frustrated over the fact that they still had not been allowed to occupy the area in Anatolia which had been promised to them in the secret treaties. They decided to act on their own and from mid-March until early May Italian troops were landed at Adalia. The Italian government tried to justify its action by claiming that the Italian troops had moved in to suppress disorder at the request of the local Turks[54] and in accordance with the article VII of the Mudros Armistice which stated 'The Allies have the right to occupy any strategic points in the event of a situation arising which threatens the security of the Allies.'[55]

On 2 May Italy sent several ships to Smyrna. By 5 May it became obvious that Italians had already started to occupy the territory they regarded as rightfully theirs.[56] This increased the anti-Italian sentiment among the Big Three (Wilson, Lloyd George and Clemenceau) and created a common concern against Italian schemes in the region. As a result it was agreed that an Allied ship representing all the parties should be sent to Smyrna at once. The British admiral, Calthorpe, was appointed to the command of the Allied ship for the landing procedure. It was also decided that the initial landing party should compose entirely of French troops, which would in turn be relieved by Greek forces. According to a later British Foreign Office report 'This mandate [to Greece] was chiefly given at that moment in order to forestall an Italian occupation on Smyrna'.[57] The landing took place on 15 May.[58] The French, Italian, and British contingents met no opposition. But when the 15,000 Greek troops landed and marched through the Konak district of Smyrna demonstrations broke out, marking the first spark of the Turkish patriotic resistance all over Anatolia.[59]

Greek soldiers in the square at Konak started firing on the barracks where Turkish troops were confined. After being under heavy fire for over half an hour the Turkish officers inside surrendered. The Turks were then marched from the Konak square to the quay where they were put in a prison ship. Allied officers on board their warships witnessed the proceedings. According to their accounts many Turkish officers failed to reach the prison ship and were knocked down, bayoneted or shot. When the news spread great demonstrations broke out in the Turkish quarters of the city. Some Turks succeeded in procuring arms from military depots and resisted the Greek troops. At the end of the first day there were more than 400 Turkish and 100 Greek casualties, killed or seriously wounded. Soon killing and looting spread through the town into the neighbouring villages.[60]

The Greek occupation of Smyrna was the most important single issue in the region to face the Allies since the end of the war. Prior to the landing both the British War Office and the Foreign Office protested at the decision and pointed out the possible consequences of such an act. Churchill was adamant and commented later:

> I cannot understand to this day how these eminent statesmen in Paris, Wilson, Lloyd George, Clemenceau and Venizelos, whose wisdom and prudence and address had raised them under the severest tests so much above their fellows, could have been betrayed into so rash and fatal step.[61]

Although the Greek occupation checked Italian aspirations on the Aegean seaboard, the Allies came no closer to the desired long-term settlement in the region. Now that the Greeks were tied up in Smyrna and Thrace, the question remained as to how to find further additional troops to calm the rising nationalist tide from Constantinople, Thrace, the Aegean coast to the far away Black Sea regions of

Anatolia. The troops sent to secure British interests, themselves aggravated the problem to an unprecedented level and extra military support was called for.

Resistance in Anatolia

The Greek invasion of Smyrna precipitated a process which was not foreseen in the scheme of things laid out on the conference desks in Paris. Churchill was right to call it a fatal step as it ignited national resistance more effectively than any strategy the Turks could themselves have organised.[62] There had been some resistance groups as early as autumn of 1918. Local 'Associations for the Defence of Rights' had sprung up all over Anatolia and Thrace as a result of the initiatives of local land-owners, teachers, craftsmen and religious notables – i.e. the provincial elite.[63] Prior to the landing in Smyrna these low-profile local groups had campaigned for certain nationalist demands but they refrained from armed aggression.[64] The President Wilson's 14 points had aroused the hope of a fair settlement.[65] They especially pinned their hopes on the twelfth which reads as follows: 'The Turkish portions of the present Ottoman Empire should be assured a secure sovereignty, … .'[66]

After the Greek landing in Smyrna, however, all optimism vanished and was replaced by the pessimism of partition. The invasion of Smyrna by the Greek forces coincided with a rise of national sentiments against a perceived anti-Turkish conspiracy. Large public demonstrations were organised in Constantinople as well as in Anatolia. *Karakol,* the Turkish underground resistance organisation, was instrumental in most of these demonstrations. *Karakol* was led by Kara Kemal, the former minister of supplies, and organised in Constantinople among the former members of the Committee of Union and Progress.

The rising tide of Turkish resistance, ranging from small gatherings in distant villages to mass demonstrations in big cities, can be traced in a number of memoirs written by the established political figures of the period 1918–23.[67] This information was also confirmed by Soviet documents; for instance, Zaydel, the Soviet representative in Odessa, in his report to Moscow, provides a detailed description of the early resistance in Anatolia.[68] On the basis of already existing local support groups an organised resistance movement took off from mid-1919 onwards, finding a leader in Mustafa Kemal Pasha.[69] Mustafa Kemal had been an unknown officer until he had achieved national fame through his role in the successful defence of

Dardanelles against the British forces in 1915.[70] During the rest of the war he had served with distinction on the Eastern and on the Syrian fronts. In the winter of 1917–18 he had made a visit to Germany as honorary aide-de-camp of the crown prince, Vahideddin.[71] Throughout his career he had avoided close association with the politics of the Young Turks, officially known as the CUP. He was publicly known to oppose the leadership of the CUP and managed to keep his name free of the discredit which was attached to the mismanagement of the war effort.

Kemal returned from the Syrian front to the capital on 13 November 1918, the same day that the Allied fleet entered the Bosphorus. He spent the winter of 1918–19 in Constantinople vying for a place in the new cabinet. During this period he was reported to have attempted to make contacts with the British. In an interview with G. W. Price, the correspondent of the *Daily Mail* in Constantinople, Kemal made the following statement:

> If the British are going to assume responsibility for Anatolia they will need the co-operation of experienced Turkish governors to work under them. What I want to know … is the proper quarter to which I can offer my services in that capacity.

Price says he gave an account of this conversation to Colonel Heywood, the senior general staff officer of the Intelligence Branch of the Salonica army. Heywood dismissed it as unimportant – adding that 'there will be a lot of these Turkish generals looking for jobs before long.'[72] For obvious reasons Kemal does not make any reference to this meeting with Ward Price in his quite detailed memoirs, *Nutuk* [Speech], a chronological description of the Turkish national struggle under his leadership.[73] What is clear in his own account is that he was advocating a political solution with the aim of influencing the Ottoman parliament as well as the British administration in Constantinople that would minimise the damage to the defeated Empire. His ultimate objective was to prevent the partition of the Turkish state.[74] He was not alone in this approach. A number of leading officers of the Ottoman army were trying to stimulate a persistent but low profile campaign through personal contacts and lobbying.

Ali Fuat Pasha, a classmate of Kemal at the Harbiye (Military Academy) – then the head of the Twentieth Army Corps in Ankara – describes in his memoirs that he and Kemal discussed the possible ways for national salvation in Constantinople in late 1918. They agreed on a programme of six points: 1) halting the demobilisation;

2) halting the surrender of arms and ammunition which were necessary for the protection of the Turkish lands; 3) securing the appointment of young and able officers to command the troops in Anatolia; 4) trying to make sure that reliable nationalist civilian administrators would remain at their posts; 5) preventing party strife in the provinces; 6) improving the morale of the population.[75]

Another leading officer of the Ottoman army, Kazim Karabekir Pasha, refers to similar talks in his memoirs. He had been the chief of staff to the German Marshal von der Goltz, who was the commander-in-chief of the Turkish armies which had defended Baghdad against the British forces under the command of General Thomson. After the armistice Karabekir returned to Constantinople. Later in March 1919 he was appointed as the commanding officer to the Ottoman troops on the Eastern/Caucasus front where sizeable forces were concentrated.

Karabekir was to command the Ninth Army that was created in 1918 and consisted of only four divisions. However, these were on full strength and well-equipped, a total of 30,000 men with a supplement of about 20,000 militia and gendarmes. Perhaps most important of all, in contrast to the troops on the other fronts, the morale was high. While other Ottoman troops were retreating during 1918, it was the Ninth Army that had actually moved forward and reached the shores of the Caspian Sea, occupying large areas in Armenia, Azerbaijan and northern Persia. After the armistice the troops had been ordered back from Azerbaijan and garrisoned around the borders of Russian Armenia, between Trabzon and Erzurum. Now Karabekir became the commander of this prestigious Ottoman army. As a formality to satisfy the Allies, the Ninth Army was reduced to the status of an Army Corps and renamed as the Fifteenth Army Corps. It made no practical difference to its actual military strength and position. This unit was destined to be the military backbone of the Turkish resistance movement in the coming months.[76]

Before leaving the capital for his post in Erzurum, Karabekir visited Kemal in his house in Sisli, on 23 March 1919. Karabekir states in his memoirs that during this visit he urged Kemal to join him in Anatolia. Karabekir writes that Kemal declined this offer as he was still determined to work for a political settlement through the parliament in Constantinople and was trying to secure a ministerial position to achieve his aim.[77]

Kemal, relentlessly energetic but still relatively inconspicuous, did not get a position in the cabinet;[78] but on 5 May he was appointed by the Ottoman government as inspector general of the newly organised Ninth Army based in eastern Anatolia. Kemal claims in his *Nutuk* that his appointment to this post came as a form of banishment because of his nationalist activities in the capital.[79] However, it would be highly unlikely that a suspect officer would be appointed to such a sensitive post with almost unlimited powers. If he had been considered a dissident it would have been much easier for the Turkish authorities in Constantinople, under the scrutiny of the British control, to put him in jail like many other nationalist activists. A total of 107 people, including a number of leading politicians and officers, were arrested in the capital on the basis of anti-government and anti-Allies activities between January and April 1919.[80]

Kemal's appointment was in fact prompted by the Allies' demand to the Ottoman government to stop the harassment of the Christian villages in the province of Samsun by local Muslim bands. Samsun is on the Anatolian coast of the Black Sea, about 400 miles east of the Bosphorus and 300 miles west of Batum. The inspector general was to be basically responsible for peacekeeping and demobilisation. Kemal in person was chosen probably because he had prestige with the armed forces and was, at the same time, acceptable to the non-CUP circles in the Sublime Porte as well as to the Allies due to his history of opposition to the CUP's pro-German war-time policies.

Soviet documents claim that Mustafa Kemal was sent to Anatolia by the pro-British government in Constantinople to check the rapid growth of the resistance movement which had just started and was felt to be a threat to the authority of Sublime Porte by the end of spring 1919.[81] It was also quite probable that Kemal's personal relationship with the monarch played a part in this appointment. Vahideddin (with whom Kemal had gone to Germany in late 1917 as his honorary aide-de-camp) had succeeded to the throne on the death of his elder brother Reshad on 3 July 1918.[82] A number of accounts state that Kemal had several audiences with the Sultan Mehmed VI Vahideddin immediately prior to his appointment.[83]

Whatever the reasoning behind this appointment, the dynamics of the circumstances around it forced Kemal into a quite extraordinary position. It provided him with an opportunity of playing a unique role in the future of the country by bringing him in close contact with the already existing local resistance groups in Anatolia. Four days after the Greeks entered Smyrna, Kemal landed at Samsun on 19 May 1919 and found himself in the midst of hectic nationalist activity. Arrangements were under way for the co-ordination of scattered resistance groups and the creation of a centralised resistance movement.[84] Samsun was at a comfortably long distance away from the Sublime Porte and the British, and it did not take long for Kemal to commit himself to organise, develop and lead nation-wide resistance.

In the following months, the events in Turkey gained an unexpected momentum and reached a crisis stage by late 1919. By then the resistance movement developed into a full war of independence 'giving a cause for apprehension' to the Allies[85]. The initial optimism of the British about the conduct of the Turkish settlement was gradually eroded by the realisation that it might be after all a long and costly process. By late 1919 a smooth settlement to the Turkish problem was no longer within the reach of the British government.[86] It appeared that 'the sick man of Europe' was not ready for his deathbed. In Nicolson's words, 'the victim recovered overnight'.[87]

British Troops in the Caucasus

Trouble for Britain was not confined to Turkish affairs. There were even more serious difficulties in the Caucasus. The months following the armistice proved that the task of securing British interests in the region would not be any easier than that in Turkey. The reason for this, however, did not stem from any local resistance as in Turkey's case. It mainly related to the complicated maze of regional problems which the British soon found themselves to be deeply involved in. An effective system of administration could perhaps have established relatively harmonious control had the British been prepared to undertake such a responsibility.

Following the Turkish armistice British troops duly moved in and put all key points under effective military control without facing any resistance. Thomson and Forestier-Walker established their headquarters in Baku and Batum respectively, and the Batum–Baku railway was secured easily. In this way by the end of 1918 British troops formed a solid 'cordon' across the Caucasus from the Black Sea to the Caspian. This was supported by an impressive fleet in the Black Sea and a small but significant presence in the Caspian.[88]

The Transcaucasian republics, Armenia, Azerbaijan and Georgia, had all declared their independence in May 1918. By September of the same year all had come under at least partial occupation of Turkish and German armies. Following the end of the war, they had the opportunity to claim back their independence. In December, after the arrival of the British in Azerbaijan, Khan Khoiski set up a new, much less pro-Turkish government with the participation of all ethnic and political groupings. In Georgia a Social-Democratic government replaced the Chkhenkeli's pro-German cabinet which had been set up under the German protectorate. In Erevan, the Dashnak-dominated Armenian government welcomed the British decision to take control of strategic points in Transcaucasia and entered into direct relations with the British representatives.[89]

The British army was enforcing the armistice and preserving the peace. It was considered as Britain's prerogative. After all, it was the threat of Turkish invasion which had prompted Dunsterville's Caucasian 'adventure' in August–September 1918. With the defeat of Turkey it was considered that Britain was entitled to the dominant position in the regions of the Caucasus because it was the British who snatched back these lands from the Turks.[90]

But the situation in the region at the end of the war was simply chaotic. The withdrawal of the Turks from Azerbaijan had left the country with no effective control, and preserving peace demanded more involvement from the British than they were willing to give. The relations between Georgia, Armenia and Azerbaijan were far from harmonious, mainly because of their long-standing territorial aspirations. After the armistice the easing of the external pressures permitted the hostilities among the independent republics to resurface. As soon as the Turkish armies left the region an armed struggle broke out between Armenia and Georgia over Borchalo region which both sides claimed as theirs. The British military leaders, who had just arrived in the region, had to intervene and use force to gain control.[91]

The boundaries between Azerbaijan and Armenia proved to be a more serious source of conflict. Many Azeri and Armenian villages were located side by side and often used the same common lands for cattle and sheep grazing. Both sides claimed such districts as theirs. Zangezur, Nakhichevan and Karabakh districts were major trouble spots.[92] These kinds of troubles provided a constant source of ethnic clashes, seriously damaging the stability in the region. In addition to such ethnic and territorial clashes there was the issue of conflicting interests of Denikin's Volunteer Army and the Transcaucasian republics. Denikin, who had the idea of 'Great Russia, One and Indivisible', had declared that he would not recognise the independence of the Transcaucasian republics.[93] When in early 1919 Denikin moved southward from the Kuban region, his armies came closer to the borders of independent Azerbaijan and Georgia. The British commanders in the region found themselves in between two opposing parties, both of whom were supported by the British government. The invasion of the Sochi district of Georgia by the Volunteer Army in January brought these two sides to the verge of a full-scale war which was prevented only when the British army intervened.[94] The task of maintaining order while securing British interests was a very tough job in such a precarious area.

Perhaps clear directives from home would have eased the situation. But none were forthcoming. In fact for London the involvement of the British troops in the regional disputes constituted an additional source of confusion to the already unsettled state of British policy. The war victory over Turkey more or less automatically initiated the British military control of the region but did not bring any clarity to what the British troops would do in the region in long-term. The future of the Caucasus was essentially a Russian question while closely linked to the Turkish settlement. As far as the Russian problem was concerned, there appeared to be no coherent policy.

The armistices between the Allies and Central Powers, which ended the war, changed the considerations of the British government. During the year between the Bolsheviks' seizure of power and the end of the war Allied intervention in Russia had initially been justified by reference to strategic 'necessities' of the war with Germany and the advance of the Turkish troops in the Caucasus.[95] The British government had been supporting the anti-Bolshevik White Russian armies in Russia since as early as December 1917 and the British soldiers had been in Russia since March 1918 when they had landed at Murmansk together with their Allies.[96] The military intervention had initially been defended as an effort to re-establish the Allies' Eastern front against the Central Powers. When the war was officially over the wartime rationale for military intervention was no longer valid. The original anti-German, anti-Turk objective behind the Allied expeditions in the Caucasus and Russia lost its significance. The British and their allies were then faced with the task of redefining their role in Russian affairs.

As the Army of the Black Sea moved in various strategic parts of the Caucasus, a series of discussions started in London. At first an initial opinion similar to the attitude in Turkish affairs appeared. At the meetings of 13–14 November 1918, it was decided to reaffirm the British government's adherence to the Anglo–French convention of 23 December 1917.[97] According to this the British were to take charge of operations 'against the Turks' in the 'Cossack territories, Armenia, the Caucasus, Georgia, and Kurdistan'.[98] But this decision alone did not clarify the general policy. There was still nothing decided as to the future role of the British army there.

On 5 December 1918, H. A. L. Fisher, historian and at the time Lloyd George's education minister, was writing in a memorandum:

> There is no doubt whatever that the continuance of our military operations in Russia, now that an Armistice has been signed with the Central Powers and their allies, is extremely unpopular with the working men and women of this country ...
>
> Probably only a small minority of working men in this country sympathise with the Bolshevik regime, but a very much larger number consider that the constitution of the Russian government is an affair to the Russians themselves and that Russia should be left to stew in her own juice.[99]

Both the Eastern Committee and the Cabinet spent most of their time in December for endless debates on the Russian policy and the Caucasus. On 23 December Lloyd George quoted Mr Fisher in the Imperial War Cabinet and stated that 'all over the country the question was repeatedly asked why the government was interfering in Russia.'[100] By the end of 1918 no definite conclusions were agreed to. The general trend of the discussions[101] was as follows: British forces should not be withdrawn from the Caucasus until after the Turkish and German forces had been withdrawn completely; a second British division should not be sent to the Caucasus without the authority of the cabinet; and British forces should not be maintained there any longer than could be avoided. The only agreement seemed to be the determination to get the troops out as early as possible.[102]

The Decision to Withdraw

On 2 January 1919, almost two months after the British forces had established firm control in Transcaucasia and in Turkey, the British Foreign Office drew up the proposal that the fighting in Russia should end at once. A meeting was proposed to establish the conditions of a permanent settlement, and the Soviet government at Moscow, General Kolchak at Omsk,[103] General Denikin at Ekaterinodar,[104] M. Tchaikovsky at Archangel,[105] and the governments of ex-Russian states would be invited.[106]

Stephen Pichon, the French minister for foreign affairs and one of France's five plenipotentiary delegates to the peace conference, immediately rejected the idea of a Russian conference which would include the Bolsheviks. He argued that the proposal meant the recognition of the Bolshevik regime.[107] The Italian delegation at the Peace Conference joined this opposition. As a result it was decided that the Bolsheviks would not be invited to Paris. It was also agreed that none of the various anti-Bolshevik factions would be given an official hearing.[108]

Even after the conference had decided not to receive the Bolshevik and anti-Bolshevik representatives officially, Lloyd George continued to advocate his initial proposal. On 20 January Lloyd George met the British Empire delegation and stated that unless some efforts were made to bring together the fighting parties within Russia, the British government should immediately withdraw its own troops and stop subsidising the others.[109] There were, in any case, not enough troops to control the regions of the Caucasus, the Ottoman capital, together with the Straits at the same time. He was, apparently, in favour of consolidating the positions in Turkey at the expense of the Caucasus.

Lloyd George's proposals were formally accepted by the Empire delegation.[110] The next day, 21 January, at the meeting of the council, President Wilson provided further support to Lloyd George's position by reporting of a series of conversations which had taken place in Stockholm between Maxim Litvinov, Soviet plenipotentiary representative, and W. H. Buckler, a US State Department official. According to Wilson's report the Soviet government was prepared to compromise on all points – which included the protection of the existing foreign enterprises, the granting of new concessions in Russia, and the settlement of the foreign debt.[111]

In the afternoon session of the council, on the same day, both Lloyd George and President Wilson initiated the idea of a meeting at Salonica, Limnos or any other location in the region as an alternative to inviting the various Russian factions to Paris. President Wilson stated that 'Europe and the world cannot be at peace if Russia is not.'[112] The following day, 22 January, Wilson put forward a draft proposal to the council, stating that the main purpose of the Allies with regard to the Russian question was to help the people of Russia; they did not intend to interfere in their affairs in any way. The proposal further stated that they aimed to bring peace, to restore order and help the Russian people to relieve their own distress. In order to achieve these ends, an invitation was issued to

> every organised group that is now exercising or attempting to exercise political authority anywhere in Siberia, or within the boundaries of European Russia as they stood before the war just concluded (except in Finland) to send representatives ... to the Princes Islands, Sea of Marmara, where they will be met by representatives of the Associated Powers, provided in the meantime there is a truce of arms amongst the parties invited, and that all armed forces ... shall be meanwhile withdrawn, and aggressive military action cease.[113]

The venue chosen for the proposed meeting was an interesting one. The Princes Islands are a group of nine small islands only ten nautical miles from the Sublime Porte in the Sea of Marmara, then under the control of the Allies. Prinkipo (Büyük Ada), the largest of the islands and a fashionable summer resort with a large number of attractive residences and several good hotels, was considered suitable for the proposed meeting.[114] The idea of an international conference was not completely new to the island Prinkipo as it had long proved to be a convenient place for important regional gatherings. At the same time as the proposed Russian conference was being discussed, a joint conference was being held in the Imperial Hotel on the island in January 1919 between the leaders of the Greek community and those of the Armenian community of Constantinople with the intention of co-ordinating the activities of these two minority groups in post-war Turkey.[115]

Prinkipo was chosen by the Supreme Council over Salonica or Lemnos probably because of the availability of plenty of vacant hotel space and its nearness to Russia. It was argued that travel to that place would not involve the transit of the Soviet delegates through any third country. The proposal in the form of an invitation, was broadcast to Russia by short-wave radio on 23 January.[116] The Allies promised to provide every facility for the journey of the representatives. The representatives would be expected at the island of Prinkipo by 15 February. The Allied powers appointed commissioners and they awaited a reply from Russia.

When the Supreme Council's radio message was heard in Moscow, Chicherin, people's commissar for foreign affairs, telegraphed the Soviet representative Vaclav V. Vorovsky[117] in Stockholm and asked his opinion about the real implications of the proposal. Chicherin held the view that if the Allies really desired to bring peace to Russia the only way to achieve it was to cease their direct involvement and military intervention in the internal struggle of Russia. Chicherin also added that he did not see how a conference on the lines suggested could bring peace to Russia.[118]

Lenin first approached the proposal with a similar suspicion. In a telegram to Trotsky dated 24 January he wrote that 'I am afraid that he [Wilson] wants to establish his claim to Siberia and a part of the south, having otherwise scarcely a hope of retaining anything'.[119] However, he also asked Trotsky that 'the person to visit Wilson will ... have to be you'.[120] Trotsky promptly refused this idea.[121]

On 28 January Chicherin sent a radio message to President Wilson. This stated that Moscow had only heard of the forthcoming Prinkipo conference from radio broadcasts and that his government would reply only when the invitation was forwarded directly to the Soviet government.[122] Both Lloyd George and President Wilson were, however, reluctant to issue an official invitation to the Bolshevik government as it might be considered a premature recognition of the Bolshevik regime.[123]

Whether receiving an official invitation or not, the Prinkipo conference was considered as a convenient 'breathing space' which might relieve some of the pressure upon the Bolsheviks. The longed-for world revolution had not occurred. It was a difficult time for the Soviet regime which was struggling for survival and trying not to yield under the heavy pressure of domestic and external problems. A serious danger of famine and lack of fuel reserves for urban centres, coupled with a run-down transportation system, were threatening the unstable power base of the Bolsheviks. On the other hand, they were faced with the attacks from White Russian armies and separatists militarily supported by the Western Allies on different fronts across the vast territories of the former Russian Empire.

During the winter of 1918–19 Lenin was afraid that the end of the war would enable the Western powers to turn their attention to Russia and destroy the Soviet regime. A 'breathing space' was desperately needed. To achieve this goal the Bolshevik leaders had initiated a persistent 'peace offensive'. From August 1918 through January 1919 the Soviet government had officially proposed peace to the Western powers on at least seven different occasions.[124] Lenin in his report to the Sixth Congress of Soviets on 8 November 1918, stated that to try to avoid war was the most urgent duty of the day.[125] He was explicit in his acknowledgement, stating that 'we are weaker … than international imperialism … [we] must do everything we can to avoid a clash with [it]'[126]

The rapid change in the international situation at the end of the war, particularly British control in Turkey, had been the main source of his concern from late-1918 onwards. Lenin was apprehensive that following the armistice agreement with Turkey, 'the British troops are ready to attack Russia from the south, through the Dardanelles.' He had also anticipated the fact that, due to their strongholds in Turkey, the British would take the Baku region under their control with the intention of strangling the Soviet regime 'by depriving us [the Bolsheviks] of raw materials.'[127] As the Western powers 'are closing in around the Soviet Republic' Lenin urged to make every effort to stop or relieve the imperialist aggression with bids of compromise and negotiation.[128]

This theme was repeated during the First Congress of the Communist International,[129] where the 'working masses of all countries' were called upon to press their governments by all available means for the withdrawal of armies from Russia, for the 'abandonment of any direct or indirect policy of intervention' in Russian affairs, the 'recognition of the Soviet government', the 're-establishment of diplomatic relations' and the 'resumption of trade relations' with the Soviet regime;[130] the protection of the 'living example', the Soviet Republic, was 'more effective than manifestos and conferences'. 'National interest' and 'world revolution' were defined in such a way as to make them virtually indistinguishable and as inseparable as the two parts of the same whole.[131]

Under these considerations the Prinkipo proposal was regarded as an opportunity towards the realisation of the short-term objectives of the regime: to 'buy time' which would enable Soviet Russia to regroup its forces and to consolidate its internal position. The Soviet government finally responded positively and declared on 4 February that it would enter into negotiations at Prinkipo.[132]

Apart from three Baltic governments and the government of Soviet Ukraine, no other Russian group accepted the proposal. The Georgian government refused because it did not consider itself as part of Russia.[133] The Whites, on the other hand, were quick to see what was behind the proposal. Fearing that the West would soon cease to support them, the three most important factions of the anti-Bolshevik forces refused the Prinkipo invitation. General Denikin, the leader of the Volunteer Army in the south, had already made it clear that he would not accept any deal in his fight against the Bolsheviks.[134] Admiral Alexander Kolchak, head of the anti-Bolshevik regime in Siberia, declared that no armistice with Bolshevism was possible.[135] The North Russian government in Archangel announced that the invitation was morally unacceptable.[136]

The Russian Political Conference in Paris, representing the 'united governments of Siberia, Archangel and Southern Russia', formally refused the invitation on 12 February 1919.[137] The refusal of the Whites was strongly encouraged by the French government and by Winston Churchill. In his memoirs Churchill defends his position by saying that the 'whole idea of entering into negotiations with the Bolsheviks was abhorrent to the dominant elements of public opinion, both in Great Britain and France.'[138] Churchill assured Konstantin Nabokov in London that he would make sure that the War Office continued to provide the Whites with all the necessary supplies.[139] Admiral Kolchak confirmed this: 'While the British government advises an arrangement with the Bolsheviks, they continued to furnish me with generous supplies.'[140]

With the hope of reaching a formula acceptable to all sides, Lloyd George sent Churchill to Paris to obtain a decision from the Peace Conference. In Paris, Churchill formulated two resolutions, a renewal of the Prinkipo proposal re-worded on a less favourable lines for the Bolsheviks and a plan for the establishment of an 'Allied Council for Russian Affairs with political, economic, and military sections' responsible to the Supreme Allied Council, mainly for the co-ordination of all efforts against the Bolsheviks.[141] As the head of the War Office Churchill knew well that the British could not afford a full-fledged support to the Whites even though he would personally have preferred unlimited aid. What he tried to formulate was, therefore, a compromise or a last attempt to continue the assistance to the White armies and to provide them with a last chance to reorganise themselves with already available sources.

In the end neither Churchill's modified project nor the Prinkipo idea was put into practice. A number of factors accounted for the failure of the Prinkipo scheme.[142] First, the determination behind the proposal was weak and widely divided. The Allies were not in complete agreement on the possible benefits of the outcome. The French identified the idea as a temporal deviation of the Allied policy and they encouraged the Whites to decline the invitation.[143] Churchill advocated a similar line by indicating that the British help would continue to the White Russian armies. On the other hand the promoters of the idea, President Wilson and Lloyd George, made no consistent effort to counteract this strong opposition. They did not try to persuade the White Russians to approve the invitation. The White Russian armies were totally dependent upon Allied aid and would probably have accepted the invitation if the Allies had whole-heartedly coerced them to the idea.

The discussion on this proposal, however, served an important goal. It brought the issue of the open negotiations with the Bolshevik government to the centre of political debate. Although a failure, the basic elements which made up the Prinkipo proposition continued to influence British policies toward Russia. Soon it was followed by other initiatives.[144] This new line found a strong ally in British public opinion to withdraw from the Russian affairs.

General Sir Henry Wilson, chief of the Imperial General Staff, presented on 21 February a paper with proposals for a complete British withdrawal from Russia at the soonest possible moment. He was convinced that this was for the best because he thought that the

Allies could not jointly arrive at a satisfactory policy for coming to terms with the Russian question. The issue was too big and far beyond the capacity of Britain to handle with a contingency of token troops.[145]

The real benefit of 'Prinkipo mentality' was felt later in early March when an interdepartmental conference was summoned by the Foreign Office. This formulated three recommendations to the War Cabinet: to make all necessary preparations to withdraw troops from Russia; to provide General Denikin with material aid and munitions of war, together with a small military mission in compensation; and to make it a condition of such support that General Denikin should not interfere with the Georgians and the other independent states in the Caucasus. All these recommendations were accepted by the War Cabinet.[146] This dramatic step on the part of the British government clearly pointed to the acknowledgement of a practical necessity which was the same rationale behind the doomed Prinkipo proposal. Without embarking upon a much larger scale war, it was impossible to defeat the Bolsheviks. All the reports coming from the region indicated that 'the Bolsheviks were militarily stronger' and making further advances on the northern, eastern and western fronts, and in the Caucasus.[147]

A direct corollary of this critical decision was related to Turkish affairs. After the recent demanding commitments in Turkey even a small addition to British commitments in the Russian Civil War was quite out of question. It was realised that Britain had to make a decision over the military commitments to Turkey for the defence of the Straits and Constantinople. British military leaders and the War Office warned the government on several occasions that the reinforcement of the Turkish capital had to be at the expense of the British presence in Batum and the Caspian.[148] The same concern was later pronounced by Churchill, in August 1919: 'We have maintained a force of some 40,000 in Constantinople and on the Black Sea shores ever since the 11th of November[1918]. The strain of this upon our melting military resources is becoming insupportable.'[149]

By March 1919, less than three months after the dispatch of British troops to the Caucasus, Lloyd George's cabinet decided to withdraw from the region.[150] All troops would be called back from the Caucasus by the end of 1919, with the exception of a garrison at Batum.[151] The British forces were to confine themselves to the control of the Black Sea and the Straits, bordering the sensitive southern flank of Russia. It provided a coherent simplified framework for British strategic interests in the region and a local focus for a unified military presence.

Various opinions exist as to why the British government decided to withdraw troops from the Caucasus. John Buchan, writing in 1923, says that due to the widespread resistance among the local population to the policies of the White Russian generals, the British decided to abstain from all active intervention in the region.[152] According to Hovannisian it was the British prime minister's 'defeatism' as well as the antagonistic policies of the Whites which convinced the British to withdraw.[153] Ullman, on the other hand, puts emphasis upon the 'domestic political objections to the immense costs, both human and material', of the British involvement in Russian affairs.[154]

It is of course difficult to deny the importance of all these elements in the British government's disheartened withdrawal from further direct involvement in Russian affairs. One can also detect, however, a particular shift in British strategic priorities in the region of Near and Middle East, of which the Caucasus is an integral part. Busch agrees with this view by saying that 'the Prime Minister's need for the Caucasian troops in Turkey was decisive in producing the withdrawal.'[155] decision to withdraw troops from all over the Caucasus except Batum represents a definitive shift in focus towards Turkish affairs.

Those four months from the armistice with the Ottoman Empire in late October 1918 to the decision to evacuate the Caucasus in March 1919, proved to be a testing time for the British government. It realised that the control of vast lands of Turkey and the Caucasus at the same time would be difficult, if not impossible. The initial opportunity to establish political control in Turkey and the Caucasus gradually appeared to be turning into a fierce struggle for territorial control. The Bolshevik forces in the north-east and the recently emerged Turkish nationalists in the south-west had already been forcing their way to power through the battlefield, or building up forces and military strength to do so.

In London the initial expectation of an effortless political settlement as a bonus of the war victory was being replaced by a more realistic vision of a geographically limited but equally effective British role in the region. By March 1919 it was understood that an effective political control could only be possible when supported by a sensible military force and that there were not enough troops to ensure law and order over an area of more than 1,000 miles across.

The redefined British objective was to withdraw forces from the Caucasus and concentrate with all available troops upon Constantinople and the Straits. The British troops in the Caucasus would have reinforced those already in Turkey and the pace of the Turkish settlement could have been accelerated.[156] A powerful stronghold in Constantinople and the Straits and the Black Sea would enable Britain still to continue to exert influence in the Caucasus. This could have been achieved by supporting anti-Bolshevik Russian forces and keeping in close contact with the Transcaucasian republics. Retaining a small garrison at Batum even after the full withdrawal of the British troops from the region indicates the desire at least to maintain a visible presence. The abandonment of Batum would have literally meant, and was perceived to have been, a complete termination of British presence and control in the Caucasus.

Whatever strategic balancing accounted for the decision it was obviously a drawback for the British. It was a clear indication that the euphoria of war victory was replaced by a defensive wait-and-see role in Caucasian affairs. The ultimate aim was, however, unchanged: to maintain British influence in the Caucasus. The difference was that it was no longer through the troops on the ground but via an indirect influence exerted through a firm control of the Straits and the Black Sea. After all, the road to the Mediterranean and the Middle East passed through the Straits.

3

An Unholy Alliance: Russian Bolsheviks and Turkish Nationalists March 1919–March 1921

The emancipation of the oppressed nationalities and their self-determination will be realised only when Anglo–French imperialism is overthrown ... [Narkomindel radio message, 27 September 1919][1]

In existing circumstances, there will then be no obstacle, so far as I can see, to the Bolsheviks' obtaining full control of the Caspian, to their seizing Baku and Enzeli, overrunning Georgia and Northern Persia, and, what is especially important from the point of view of peace in Turkey, joining forces with Mustafa Kemal, whom the occupation of Constantinople will probably decide definitely to ally himself with Bolshevism. [J. M. de Robeck, British High Commissioner in Constantinople, 18 March 1920][2]

The Turkish government accepts responsibility ... to compel Georgia ... and Azerbaijan ... to enter into union with Soviet Russia, ... and is ready to undertake military operations against ... Armenia. [Mustafa Kemal, 26 April 1920][3]

First Contacts

The British decision to withdraw from the Caucasus (partly by choice, partly of necessity, and partly against their will) was a tactical political retreat from the more expensive of British objectives. Britain sought to play safe by securing its positions in the Turkish lands rather than through costly direct confrontation with the Bolsheviks. A firm-handed policy in Turkish affairs would facilitate a quick settlement and the country could be isolated from the potentially dangerous events of the Caucasus and south Russia, and perhaps could act as a positive catalyst in their affairs.

A confrontation with the Bolsheviks was, however, inevitable, not in the Caucasus but in Turkey. Britain was soon faced with the anti-British co-operation between the Russian Bolsheviks and the Turkish nationalists. The Bolsheviks observed the discussions in Paris with great concern. The British decision to withdraw from the Caucasus relieved them from some of their worries. Lenin in April 1919, at a conference of Moscow Factory Committees and Trade Unions, stated that 'the Allies had practically abandoned open military intervention in Russia's affairs ...'[4] Later in the same year, Frunze, commander of the troops in the Ukraine and Crimea,[5] summarised the real implication of the British decision of March 1919 as follows: 'the English asked the White generals to stop the war ... [Therefore] the civil war on the Eastern front was approaching its final conclusion'.[6] The Bolshevik government had for a long time been aiming at this both through diplomatic means and the might of the Red Army. However, it viewed the British control of the Straits, Constantinople, and the Black Sea (an area traditionally regarded as the most sensitive to the overall security of Russia) with increasing concern. From the time the armistice at Mudros was concluded the possibility of a powerful anti-Bolshevik attack initiated by Britain from the south through the Turkish lands had been considered.[7]

The Greek landing in Smyrna increased this anxiety. The Bolshevik leadership believed that Greek operations were directly controlled by the British Empire and Greek expansion in Asia Minor was thus regarded as a threat to the Soviet regime. A powerful Greece extending into Asia Minor could have blocked Soviet access to the Mediterranean in the long term. The frontiers of the Greek expansion, according to the Bolsheviks, would not necessarily stop only at the western Anatolia. In Moscow's eyes there was a danger that the Greeks would 'try to create the Great Armenian–Byzantine state' in eastern and northern Anatolia which 'could serve as a gable to hold on the fire of imperialism' on the borders of Soviet Russia.[8] In the light of these developments it was natural for the Bolshevik regime to try to establish close contacts with those elements in Turkey who were opposed to the Allied scheme in their native lands.

The first signs of an actual step to initiate relations between Turkish nationalists and the Bolsheviks can be traced back to early 1919. During the first half of 1919, a number of telegrams and reports which were sent to Chicherin from south Russia and Transcaucasia described the situation in Constantinople and Anatolia in great detail. There is a Narkomindel report dated 20 April 1919 about the severe measures taken by the Allies in Constantinople against the Turkish communists and the leading nationalists. The report states that 'in Constantinople alone the Allied police arrested

300 Turkish communists'. It continues with the observation that all such acts of repression increased anti-Western and pro-Soviet sentiment among the Turkish population.[9] The first actual contacts between the Turkish and Soviet sides were established in the Crimea, which was at the time like an isolated island between the British-controlled Caucasus and the Ukraine which had been invaded by the French. In the Crimea, a government had been established under the premiership of General Sulkevitch at the end of January 1918 under the protection of the Germans (who at that time had occupied the region).[10] Following the defeat of the Central Powers the entire situation in the Crimea was totally changed.[11] On 16 November General Koch, commander of the German troops in the Crimea, had notified the Crimean Tatar nationalists that he would no longer support their government. A new government was formed, which had established contacts with General Denikin and the Allies.[12] However, the control of the Allies and that of the White forces in the Crimea remained very limited, and the activities of the pro-Bolshevik groups continued almost intact.

As part of the Bolshevik agitation campaign, a group of Turkish communists arrived in the Crimea from Moscow on 22 January 1919. These communists, led by Mustafa Subhi,[13] were the leading members of the Centre of All Russian Muslims which had established in Moscow after the October Revolution 1917. Subhi had been organising socialist propaganda and agitation in Moscow by a Turkish paper called *Yeni Dünya* [New World].[14] He and his comrades organised a bureau of Muslim communists in the Crimea and started to co-ordinate communist activities geared towards Anatolia, and began shipping men and propaganda material across the Black Sea. When Denikin attacked the Crimea on 23 April 1919, the group moved to Odessa. The same day *Izvestiia* ran a report on the local opposition movements to the Allied presence in Anatolia and acclaimed it as 'the beginning of the first Soviet revolution in Asia.'[15] The Turkish communists arrived in Odessa in early May and from there, according to the account of Subhi, went in two separate groups to Turkey in the second half of May: one to Constantinople, the other to a Black Sea port of Anatolia.[16] This date actually coincides with Kemal's landing at Samsun on 19 May 1919. A few days later, on 25 May, Kemal went to the neighbouring town of Havza and, according to several Turkish sources, there he met a Bolshevik delegation.[17] Since there are no other details about the meeting one can only speculate on the members of the Bolshevik delegation. But as the date of the meeting quoted by the Turkish accounts coincides with the possible date of arrival of the Turkish communists as mentioned by Subhi, it seems likely that this Bolshevik delegation might be the group of Turkish communists sent from Odessa by Subhi.

On 12 June Kemal went to Amasya, an inland town 100 miles south of Samsun, from where he sent a telegram to Karabekir describing the situation and the contact with the Bolsheviks.[18] The main emphasis of Kemal's telegram was the possibility of a particular use of the antagonism between the Bolsheviks and the British. He was not proposing Soviet–Turkish co-operation at this stage, but rather emphasising the threat that such co-operation would pose to the British. From his headquarters at Erzurum, an important trading centre in north-east Anatolia 300 miles from Samsun, Karabekir sent a positive reply to Kemal. His advice was for Kemal to continue contacts with the Bolsheviks by all possible means. He was, however, very concerned not to take an open pro-Bolshevik position and advised Kemal to be cautious about British propaganda. The best approach, according to Karabekir, would be to maintain contact with the Bolsheviks at a personal and unofficial level for the time being and not to give the impression that the Turks were desperate for Soviet help.[19]

The issue of contacts with the Bolsheviks was on the agenda of the first historical meeting of Kemal with two other prominent national leaders in Amasya. At this meeting, Kemal, Rauf Orbay, former naval minister and Ottoman delegate to Mudros, Ali Fuat Cebesoy, commander of the 20th Army Corps in Ankara, and Refet Bele,[20] the highest-ranking member of Kemal's own staff, signed a protocol on 21 June.[21] The protocol, known as the *Amasya Tamimi*, was soon approved by Kazim Karabekir. With the Amasya Proclamation the principles of national resistance were drawn up in an attempt to unite 'all individual and regional efforts … into one channel'.[22]

Soviet–Turkish relations stayed on the agenda throughout the year and were discussed in much detail at the two conferences of great importance for the early stages of the national movement: the Erzurum Congress in July/August 1919 and the Sivas Congress a month later.[23] These two congresses brought together various local centres of resistance from all over the country and formulated the basic aims of national resistance as the preservation of territorial integrity and national [*milli*][24] independence. Armed action was not ruled out against the occupying forces if necessary.[25] The congresses of Erzurum and Sivas were also fundamental in preparing Mustafa Kemal's career as self-appointed leader of the national struggle. By keeping his relations balanced with different parts of the movement, and allying himself with his influential political and personal friends (of whom Karabekir and Ali Fuat were the most important), Kemal gradually emerged as an obvious choice to lead the struggle.[26] He soon established businesslike contacts with the other local and military leaders. He was appreciated for his mind, his energy and his political vigour. Kemal was elected as the chairman and directed the discussions in both Erzurum and Sivas Congresses.[27]

While the Turkish nationalist leadership was busy discussing and formalising the possibilities and the prospects of a relationship with the Bolsheviks to provide support for the resistance movement in Anatolia, another contact was established through a different channel. The main architect of this link was Enver Pasha, Turkey's flamboyant war-time leader, still influential yet somewhat quixotic. He, with the other leading Unionists, had fled from Turkey immediately after the conclusion of the armistice and ended up in Berlin among a small colony of Turkish political refugees towards the end of 1918.[28]

Enver held a number of discussions with the Soviet emissary Karl Radek, who was in prison in Berlin.[29] The main theme of these discussions was to create an anti-Entente alliance between Turkey, Soviet Russia and Germany. Several high-ranking German military officers also participated. Enver Pasha, determined to fight what he called 'the second phase of the war', identified the conflict in Turkish lands as an opportunity to create a powerful front against the Allies.[30] We know from the Turkish sources that Enver Pasha went to Moscow a year later in mid-1920 and continued his contacts there for an ambitious Soviet–German–Turkish alliance.[31] Very little is known about the results of these talks and there has been a great deal of speculation regarding the significance of the whole project.[32] But a number of Soviet documents indicate that Enver Pasha played an important role in the Bolsheviks' approach to the Anatolian movement during its first three years – 1919 to 1921.

There exists, for example, a Narkomindel letter written in April 1921 to the Central Committee of the RCP(B) which states that 'Enver has already showed great sacrifice for the realisation of the Turkish–Soviet relations.'[33] However, the most important evidence we have of this phase of Turkish–Soviet–German relations consists of a letter to the Narkomindel from Berlin, written by Major Tschunke. Fritz Tschunke, as a military expert, played a central role in the military relations among the three countries.[34] Tschunke's letter, dated 14 July 1920, states that an agreement had already been reached in Moscow according to which military and political authorities in

Germany had agreed to provide material assistance to the anti-Entente struggle in the region. The letter asks Narkomindel about the practical conditions of delivery saying that the Germans were ready to dispatch the military material through the Turkish Black Sea port of Trabzon, about halfway between Samsun and Batum, or directly in Petrograd.[35]

It is of course difficult to assess the real significance of such an agreement on the basis of Tschunke's letter.[36] It would not be very erroneous, however, on the basis of a number of other references to the case,[37] to assume that such an understanding did play an important role in Soviet–Turkish co-operation in terms of financial and military help to the Anatolian struggle. In addition to these a number of eye-witness accounts[38] emphasise the fact that German money and armaments (which had come through Russia) had a crucial impact in the Turkish national struggle.[39]

In this way connections were established through a couple of channels between two anti-British forces in the region. There was a mutually shared expediency which brought the two forces together with the common aim of undermining British designs in the region. Britain's decision to withdraw from the Caucasus came at the point when both the Russian Bolsheviks and the Turkish nationalists started to seek further means of collective action in order to strengthen their respective positions against the British power. The Caucasus, therefore, provided a convenient arena in which to put this co-operation into practice.

The Caucasus acquired a critical status in the Bolshevik strategy from mid-1919 onwards. Britain's intensified involvement in Turkey and her announced intention of withdrawing troops from Russia provided an opportunity for the Bolsheviks to consolidate their power within the Caucasian lands of the former Russian Empire and to fortify Russia's sensitive southern flank against the tightening British grip in the Black Sea and the Turkish Straits. The positive steps towards an alliance with the Turks provided an extra bonus in the Bolsheviks' Caucasian campaign, enabling the Bolsheviks to mobilise a significant part of the Turkic–Muslim population of the Caucasus to their side.

Barrier or Bridge? The Caucasus in Soviet–Turkish Relations

Because of its strategic location, the Caucasus had for a long time represented a front-line between the great powers. Situated at the cross-roads of Asia and Europe, the area has proved both a tempting object of conquest and a passageway to other regions. The Caucasus has always had a value out of all proportion to its size. It has no boundaries except for the two flanking seas: the Caspian Sea in the east and the Black Sea in the west. In the north, beyond the Caucasian mountains, it merges imperceptibly with Russia; in the south with Turkey and Persia. At no point can a line called a frontier be drawn. Rather the Caucasus at this time should be thought of as being itself a frontier. A borderland to bridge parties, as in the case of the Bolsheviks and Turks, or to bar the advance of one party, as in the case of Britain against the Bolsheviks. The Caucasus in mid-1919 again emerged as the strategic interface between the events of Russia and those in Turkey.

During the winter of 1919–20 the region witnessed a major Bolshevik campaign for a 'Sovietised' Caucasus. The Turkish nationalist cadres, sent from Anatolia, played an important role in this campaign. For the Turkish nationalists an independent Caucasus was an obstacle that would hinder the transfer of precious war materials from Russia. It was hoped that the establishment of Bolshevik control in the region would facilitate much needed material aid for the Turkish struggle for independence.

During the second half of 1919 effective control of the Caucasus was apparently shifting fast towards the Bolsheviks. This was a period when the tide was turning against the White armies in the Russian Civil War elsewhere. Denikin's Volunteer Army, after a number of major victories between June and October, met with a vigorous counter offensive in Tula, about 100 miles south of Moscow. Starting from May 1919, Kolchak's forces had also suffered considerable setbacks. During the first week of June the Bolsheviks drove back Kolchak's forces.[40] By the end of October the anti–Bolshevik forces were driven back on all fronts. Throughout the rest of the year Denikin's forces were in constant retreat. In the winter of 1919–20 a Bolshevik victory in the civil war was imminent. It suddenly seemed likely that the White armies would be defeated and the Bolsheviks would break through the Caucasus Mountains to the south. In January 1920, Kolchak was captured and executed at Irkutsk. Denikin's retreat continued during January and February 1920. It was not the end of the civil war; but it was the turning-point. The revolution had been saved. Victory went to the Bolsheviks, because they were better able to mobilise and organise support and resources available to them and to apply force effectively.

The final defeat of the White armies was anticipated by the British authorities in the early days of 1920. The British high commissioner in Tiflis, Oliver Wardrop, sent a telegram to Lord Curzon on 3 January 1920 warning the government against the defeat of Denikin and a powerful Bolshevik thrust to the Caucasus.[41] In the same telegram he also pointed out that Turkish propagandists and the armed militants were helping the Bolsheviks in the region and urged the government 'to send some extra military units to north-west Iran and north-east Turkey' to prevent ingress of Turkish agitators and armed militants. Three days later Lord Hardinge wrote that 'The Bolsheviks ... are concentrating their efforts in the East and the scene of their activities within the next few months will undoubtedly be in Transcaucasia ...'[42]

The issue came onto the agenda of a meeting at the Quai d'Orsay on 10 January of the Allied premiers and foreign ministers and the American and Japanese ambassadors. Immediately after the meeting the foreign ministers decided upon the de facto recognition of the governments of Georgia and Azerbaijan.[43] They considered the Armenian issue as part of the Turkish peace settlement, so no action was taken at this stage about Armenia. Neither were the claims of Daghestan dealt with – largely because it had no representatives in Paris but also because it was regarded as a lesser issue.[44]

All such efforts and half-hearted measures taken by the Allied statesmen were simply considered as too late by Field-Marshal Sir Henry Wilson. According to Wilson, a firm control was 'completely out of accord with the realities of British power'. He further stated that despite the fact that 'those fools in Paris only yesterday agreed to acknowledge the "de facto governments" of Azerbaijan and Georgia, it was quite true that Georgia and Azerbaijan would go Bolshevik.'[45] Sharing Wilson's impression, Churchill described the situation by saying that 'contract your commitments, select your obligations, and make a success of those to which you are able to adhere'. In Churchill's words, it would be much better 'to bring ... forces out of danger and responsibility' as it did not seem possible to resist the Bolsheviks in military terms.[46]

As the heads of the Allied governments were discussing in Paris, a number of British intelligence officers were active in the region to establish contacts with the intention of preventing further co-operation between the Bolsheviks, the local Muslims and Turkish nationalists. Soviet sources mention the attempts of 'Anglo–French' agents in the region to diminish the increasing influence of the Bolsheviks among the Caucasian Muslims.[47] A number of telegrams sent from the region to the Central Committee of the Russian Communist Party warned Moscow about the activities of 'experienced British intelligence officers' in the strategic centres of Transcaucasus and eastern Anatolia, such as Baku, Tiflis, Erzurum and Trabzon.[48] Similarly, several Turkish sources confirm that in eastern Anatolia a number of British officers tried to convince key nationalist leaders to break with the Bolsheviks and to fight against them in the Caucasus.[49] Here the activities of Colonel Rawlinson are worth mentioning.

As described in Part I, after playing a significant role in the unsuccessful defence of Baku against the Turkish armies in August–September 1918, Rawlinson had left Baku with Dunsterville and returned to London soon after. 'Having been just under three months at home', he had been sent back to the Caucasus in February 1919.[50] This time his orders were to base himself in Tiflis as a 'Special Service Officer' under the orders of General Beach, who was then in charge of British Intelligence in the Caucasus.[51] Following the War Cabinet's decision to withdraw troops from the Caucasus, Rawlinson was told to concentrate his activities in north-eastern Anatolia. The early months of 1919 had been quite busy for Rawlinson who tried to set up an intelligence network in the region between Trabzon, Kars and Erzurum where he had arrived at about the same time as the new Turkish army commander Kazim Karabekir.[52]

In Erzurum Rawlinson maintained unofficial *pourparlers* with leading nationalists and local officers. He even had a brief conversation with Mustafa Kemal who visited the town for the nationalist congress of July 1919.[53] On a number of occasions he tried to win Karabekir to the cause of building a mighty barrier in the Caucasus to confront Bolshevik armies. In his memoirs, Rawlinson does not make any direct reference to the contents of his discussions with Karabekir. He recalls the talks only by saying that 'we had … many discussions with reference to Russia and her future,…'[54] Nevertheless, we have enough material in the form of memoirs of activists from this period, including those of Karabekir, to draw conclusions about the basic framework of the Rawlinson–Karabekir talks. If these accounts are to be believed – and no one has ever questioned them – Karabekir was asked on several occasions to act in the region against the Bolsheviks.

Karabekir gives a detailed account of his meetings with Rawlinson. On 12 February 1920, Rawlinson told Karabekir that the Bolsheviks were advancing in the Caucasus with a large army which included some Muslim regiments and he asked Karabekir to mobilise his troops against the Red Army if it attempted to violate the Turkish border. According to Karabekir Rawlinson warned him by claiming that 'otherwise the British army would approach from the West and you would be crushed in between …'[55] Karabekir again met Rawlinson four days later. This time he found Rawlinson in rather a pessimistic mood about the situation in the Caucasus. The latter admitted in this meeting that there was very little to stop total Bolshevik control in the region.[56] He also offered money in return to a few reliable Turkish officers who would go to the region and collect information for the British.[57] Rawlinson called on Karabekir twice more – on 19 and 21 February. In these further talks he tried to sound more convincing by hinting that the British were ready to hand Constantinople and Smyrna over to the Turks if they stopped the Bolsheviks in the Caucasus.[58]

In Constantinople the idea of a Caucasian barrier against the Bolsheviks found some advocates among the Turkish nationalists. Rauf Bey, close friend of Mustafa Kemal and the leading nationalist leader in the Ottoman Parliament, had a meeting with the minister of war, Fevzi Pasha, on 11 February 1920 and urged him to convey a simple message to the British: 'if the British supports our independence in Turkey, we could play an important role in the Caucasus to stop the Bolsheviks.'[59] Another telegram sent from the British High Commission in Constantinople to London confirms the existence of such attempts. The telegram (written by Luke[60] on 27 May) which summarises a proposal initially put forward by one 'Fuat Bey' and apparently supported by Luke:

> Fuat Bey put forward a scheme … The scheme consists of raising the mountain tribes of the Northern Caucasus against the Bolsheviks … necessary financial support …, at the maximum, from 15,000 to 20,000 pounds sterling. The source of the money would remain secret, nor would any responsibility be incurred by whoever provided it. I submit the proposal for the favourable consideration of His Majesty's Government, and urge that the loss of any money supplied, should the scheme fail, would be negligible compared with the valuable results in the event of success.[61]

These attempts stopped short of creating an official understanding and effective co-operation for two main reasons. In the first place the Turkish national movement did not yet reflect a united and clearly defined front. It was a conglomerate of groups and factions with different backgrounds, loyalties, and leaders. Informal networks based upon family contacts, friendships and school networks were more influential than the official structure of the movement. While some leading nationalists were discussing the prospects of playing an active role in the establishment of a Caucasian barrier against the Bolsheviks, others were actively co-operating with the Bolsheviks to stop the formation of that barrier in the region.

According to a telegram sent by Victor Naneyshvili, a member of the Caucasus Regional Committee (Kavkazskii Kraevoi Komitet) and of the Russian Communist Party, to S. M. Kirov[62] on 20 October 1919 an agreement had already been reached in Baku between the Bolsheviks and the Turkish nationalists to establish a united front against the British Empire.[63] We find further confirmation of this agreement in the memoirs of Karabekir. Dr Fuat Sabit, who had been sent to the Caucasus to contact with the Bolsheviks on behalf of the Turkish nationalists, sent a communication to Karabekir on 19 October 1919 on this agreement.[64]

British attempts to convince the Turks to take an anti-Bolshevik stand in the Caucasus failed also because the British did not, and could not, offer any firm and substantial guarantee to the Turks. Karabekir, who gave the issue serious consideration and circulated the demands of Rawlinson to the other nationalist leaders, seemed ready to take a calculated risk as a soldier but first wanted to see some solid evidence and official commitments from the British. The British proposals were, however, clearly self-serving and possibly suspicious in their eagerness to foster the idea of a British–Turkish co-operation against the Bolshevism. Karabekir finally came to the conclusion that the British had neither a clear policy nor the means to set up a Caucasian barrier.[65] In his telegram to Mustafa Kemal dated 14 February 1920, Karabekir expressed his deep suspicion: 'how could the Allies talk about the formation of a Caucasian barrier when they have no military power in the region.'[66]

It appears that Rawlinson's efforts achieved the reverse of his intention, strengthening the belief on the Turkish side that, compared with the Bolsheviks, the British neither had the political commitment nor the necessary means to support their offers. After his meetings with Rawlinson, Karabekir drew his own conclusions that the strategic position of the Turks in the region was an effective weapon which could be used skilfully against the British. Without compromising his cautious approach to the Bolsheviks, Karabekir sent a number of instructions to his contacts in Azerbaijan and Daghestan and urged them not to get involved in any British conspiracy against the Bolsheviks.[67]

While the British role had already appeared to be none but that of a passive by-stander, the hectic winter months of 1919–20 witnessed increasing activities in the region where Turkish officers and leading nationalists were openly co-operating with the Bolsheviks in order to secure a safe passage – a material aid delivery corridor from the north.[68] The Turkish influence in the Caucasus did not vanish with the retreat of the Ottoman troops after the Mudros armistice. Some 3,000 former Ottoman military personnel and schoolteachers had stayed and entered Azerbaijani service. Now pro-Turkish sympathies surged again with the rise of nationalist movement in eastern Anatolia. Two close relatives of Enver, Halil Pasha and Nuri Pasha, who also had active links with the nationalist movement in Anatolia, played an important role in this.

Halil Pasha, the head of the Army Group of the East, and Nuri Pasha, the leader of the Army of Islam, had together played the central role in the Turks' capture of Baku in September 1918. Halil and Nuri had been arrested by the British following the Mudros Armistice in November the same year. Spending less than a year in captivity at different locations, Halil in Constantinople and Nuri in Batum, both had escaped on the same day – 7 August 1919.[69] Now in early 1920 (after the withdrawal of the British troops) Halil Pasha was in Baku as the representative of the Turkish nationalists to the Azerbaijani government. Nuri Pasha was trying to reorganise the Turkish irregulars, the remnants of the 'Army of Islam', dispersed in the Caucasus following the arrival of the British in the aftermath of Mudros in late 1918.[70] They both acted in this period as intermediaries between the representatives of the independent Republic of Azerbaijan and the Red Army.

Mustafa Kemal on 6 February 1920, in a telegram to Karabekir, elaborated how the Western powers had plotted to separate the Turks from the Bolsheviks by upholding Armenia, Georgia, and Azerbaijan. Kemal further stated that the Caucasus was 'the only possible front' where the Turkish national movement could initiate a counter attack against the Allied powers. If three Transcaucasian states, Kemal continued, could constitute a pro-Western barrier, this would be the end of all hopes for the establishment of an independent Turkish state. Therefore, Kemal urged, the Turks should do everything in the Caucasus to facilitate the Bolshevik control of the region.[71]

This support from Ankara provided an extra boost to the Bolsheviks' Caucasian campaign, drawing a significant part of the Turkic–Muslim population of the region to the side of the Bolsheviks. At this juncture, the nationality policy of the Bolsheviks displayed flexibility and rejected rigid application of class warfare in the non-Russian regions. They vigorously pursued a policy based on an appeal to the religious and national feelings of the local population by calling for creation of independent Soviet republics. The co-operation with the Turks was well utilised in Bolshevik propaganda which declared its aim as to free oppressed nations from enslavement and keep particular religious practices, customs and local traditions completely intact.[72]

This policy was first implemented in the events of the north Caucasus and Daghestan. The Provisional Terek–Daghestan government, which had been proclaimed in December 1917, could not establish an authority even in its capital Vladikavkaz and had dissolved itself in January 1918. A pro-Bolshevik Soviet regime, called the 'Terek People's Republic', had been proclaimed in Vladikavkaz in March 1918. The life of the Terek People's Soviet, however, was not longer than nine months – it was overwhelmed by the armies of General Denikin in early 1919.[73]

In late 1919, as the White armies of Denikin were retreating, the Bolsheviks and the local Muslims (supported by the Kemalist militants) increased their activities for the re-establishment of Soviet power in the north Caucasus. A report of the Communist Party of Daghestan, dated September 1919, confirms that 'the Turkish officers are helping the Bolsheviks in Daghestan. They want to diminish the British influence in the region, which they consider is essential for their anti-imperialist struggle ...'[74]

The central figure here was Grigory (Sergo) Konstantinovitch Ordzhonikidze, the political leader of the Bolsheviks in the north Caucasus and Transcaucasia. He was an old friend of both Lenin and Stalin. Ordzhonikidze had been a member of the first Collegium of the first Soviet Political Police, Cheka (All-Russian Extraordinary Committee for combating counter-revolution, speculation, sabotage, and misconduct in office).[75] When the Eleventh Red Army had been forced to retreat early in January 1919, having waged fierce battles in the north Caucasus with the forces of Denikin's Volunteer Army, the main units of the Eleventh Army escaped into the Kalmyk steppes and to Astrakhan. The other units, headed by Ordzhonikidze, continued to wage a guerrilla war in the region against the White armies. In the mountain villages of Daghestan and Chechnia Ordzhonikidze formed partisan detachments consisting of local Chechens, Ingush and Ossetians. For the rest of the year Chechnia and Daghestan remained the main centres of Bolshevik activity in the region. The detachments of the mountain partisans organised frequent raids on the rear sections of Denikin's army.[76] When Ordzhonikidze took refuge in the mountains of Chechnia and Ingushetia, a Chechen peasant, Ibrahim Geldegen, served as his bodyguard. Because of his courage Ordzhonikidze nicknamed him 'Zelimkhan of Geldegen'.[77]

During this period the Bolsheviks often made use of local nationalist leaders. They pursued an opportunistic policy based on an appeal to the national sentiments prevailing among the Muslims of the Caucasus. This policy had already been elaborated in the Caucasus Regional Committee meeting of the RCP(B) in Baku at the end of May 1919. The Committee, after a long discussion, agreed that the next step forward was to overthrow the governments of the existing republics and to establish Soviet power within their boundaries. In other words, the objective was described as transforming the 'bourgeois' republics into independent Soviet Republics with a strong emphasis upon their cultural, religious and ethnic autonomy.[78]

During the winter of 1919–20, it became clear that Soviet Russia would not only survive the civil war but emerge with most of the former tsarist empire intact. In the Caucasus it was apparent that the tide had already turned for the benefit of the Bolsheviks who were enjoying the full support of the Turkish nationalists. The British, who had already withdrawn their troops from the region, were nervous but passive by-standers of the Caucasian affairs.

The Allied Occupation of Constantinople

The same winter months of 1919–20, when British fortunes in the Caucasus were at a low ebb and things were beginning to look up for the Bolsheviks, witnessed the emergence of quite serious opposition to British control in Turkey. The profile of the Turkish national movement was rising rapidly in both Anatolia and Constantinople. The political prestige and military capacity of the Turkish nationalists were considerably boosted as a result of antagonistic operations of the Greek army in western Anatolia and the failure of the British policy in the Caucasus. The close co-operation with the Bolsheviks played a significant role in reinforcing the self-confidence of the Turks.

In December 1919, new elections were held in Thrace and Anatolia for the Ottoman Parliament and an overwhelming nationalist majority was returned to the parliament which assembled in Constantinople on 12 January 1920. On 28 January the deputies approved the National Pact [*Misak-i Milli*], based on the declarations of Erzurum and Sivas Congresses. The principal points of the National Pact had first emerged as a consensus between Mustafa

Kemal, Kazim Karabekir and other leading nationalists at the end of the Erzurum Congress in August 1919. The National Pact consisted of six articles and affirmed the unity and independence of those territories in Anatolia and Thrace 'inhabited by an Ottoman Muslim majority'. It is important to note that it advocated not Turkish national sovereignty but that of all Muslim Ottomans. In practice this meant Turks, Kurds, Lazes and Circassians of Turkey.[79]

Parallel to this development in the parliament, the nationalist militants in Constantinople stepped up their activities by organising raids on Allied arms depots and shipping the material accumulated in this way to Anatolia.[80] The British high commissioner, Admiral de Robeck, found it difficult to maintain law and order in the face of such activities. Finally in February he admitted that 'nationalism [Turkish] was ready openly to claim its own in Constantinople.'[81]

The British were convinced that the Bolsheviks were to blame for all such disturbances. A number of British intelligence reports informed the government of the role of the Bolshevik agents in Turkey and intimate co-operation between these two 'anti-Entente and especially anti-British' movements.[82] The increasing nationalist activities in Turkey were considered to be directly related to the Bolsheviks' campaign in the Caucasus. The reports regarding the presence of the Turkish officers working hand-in-hand with the Bolsheviks in the region, especially in Azerbaijan and Daghestan, led to pessimistic projections.[83] Numerous reports sent from the Mediterranean and Turkey reflect the emphasis British official representatives and the military commanders put upon the close link between the events happening in Turkey and those in the Caucasus and the implications for the future of British control in the region.[84] In March Vice-Admiral de Robeck in his report to the secretary of the Admiralty envisaged that 'in existing circumstances, there will … be no obstacle, … to the Bolsheviks obtaining full control of the Caspian, to their seizing Baku and Enzeli, overrunning Georgia and Northern Persia, and … joining forces with Mustafa Kemal …'[85]

A number of options were put forward by the British in February and March 1920 to prevent the situation from deteriorating any further. A military solution to stop the Bolsheviks in the Caucasus had already been 'reluctantly dropped' – partly because of the 'difficulty in finding the necessary troops'.[86] There was also anxiety about the security of British positions in Turkey. Under such circumstances it was considered a necessity to call more troops to the Ottoman capital to ensure the security of the existing British forces.[87]

In a War Office dispatch of 10 March GHQ Egypt demanded the provision of 'immediate military and naval reinforcements' to Constantinople.[88] News about nationalist activities in Anatolia further increased the concern about the security of the Allied positions in Constantinople. Heavy fighting between the local nationalist forces and the French in Maras, in the south-eastern part of Anatolia called Cilicia, ended with the evacuation of the region by the French and created extra panic among the Allied circles.[89] All reports dispatched from the region confirmed that the British forces in the Turkish capital were not adequate enough to stop a possible Turkish–Bolshevik attack on Constantinople. On 13 March 1920 W. S. Edmonds of the Foreign Office proposed the transfer of the 'British officers who are now with General Denikin' to the Turkish capital.[90]

Finally, the Allied Supreme Council decided on a reinforced occupation of Constantinople and a firm control over the Straits.[91] On 16 March 1920 the Allied corps under the orders of Lieutenant General Sir Henry Wilson entered the Turkish quarters of the town and arrested leading Kemalists – together with other suspected nationalists and communist activists. Those arrested were deported to Malta.[92] The Kemalists in Anatolia retaliated by arresting all British officers and personnel in the provinces under their control. Colonel Rawlinson was amongst those arrested in Erzurum.[93]

The Allied transformation into a hostile presence in Constantinople and the round up of prominent Turkish politicians represents the beginning of the next crucial phase in the course of the Turkish national struggle. It had a decisive and convincing impact on the Turks by proving that there would be no softening of the Allied attitude towards Turkey. Their response was equally momentous: the convocation in Ankara of a Grand National Assembly which established a provisional government on 23 April 1920.[94]

On 19 March, just three days after Constantinople had been occupied, Mustafa Kemal issued a communiqué to all *vilayets*, independent *sanjacks*, and the officers commanding Army Corps. The communiqué reads:

> With the object of considering and carrying out the best way to secure the inviolability of the capital, the independence of the nation, and the liberation of the country under these conditions, it has been deemed absolutely necessary to convene an assembly to be held at Ankara that will be furnished with extraordinary powers and will permit those members of the chamber that has been dissolved to come to Ankara to take part in it.[95]

Ankara was chosen because of its central location and because it was at the head of a railway line directly linked to Constantinople. In 1920 Ankara was not more than a small impoverished Anatolian town of 20,000.[96] The town, nestled into a hill some 3,000 ft above sea level, had lived through 2,000 years of recorded history. The Hittites, the Galatians, the Byzantine Greeks and the Ottoman Turks had successively made a battleground of the Ankara plateau. During the last century, the centre of activity in Ankara had shifted from the citadel high on its rock to the marshy valley below. In that valley the German engineers who built the Baghdadbahn had marked out the terminal of a spur line from the railways junction at Eskisehir, and a telegraph office (always an important part of the Sultan's espionage network) had been built adjacent to the station. A few more nineteenth-century German-style buildings had been added before the war. This 'primitive, hilly, muddy town of the interior' was to be the heart of the new Turkey.[97]

The Soviet government followed the events in Constantinople with great interest. On the day of the occupation the Narkomindel immediately put forward an announcement listing the names of those arrested during the Allied occupation.[98] The expectations of the Soviet side rose to new heights for the prospect of an even more intimate co-operation with the Turkish nationalists under these circumstances.

The expectations of the Bolsheviks were not unfounded. Kemal, as soon as he heard about the occupation of the Ottoman capital, instructed Karabekir to initiate the necessary arrangements to establish an immediate contact with the Bolsheviks for an united front in the Caucasus.[99] The recent events in Constantinople were enough to overcome Karabekir's initial caution. As soon as Kemal's telegram reached him, Karabekir sent a brief on recent developments in Constantinople to Halil Pasha (who was in Baku) and urged him to mobilise all those under his power to support a Bolshevik offensive in the Caucasus.[100]

The first official letter from the Ankara government (dated 26 April 1920) was dispatched to Moscow in this atmosphere. This letter[101] was a clear indication of the full commitment of the Turkish side. Kemalists were convinced that only co-operation with the Soviet government would provide the Turks vital material and diplomatic support at this stage. The letter, signed by Mustafa Kemal, starts with the consent of the Ankara government 'to join the fight of Soviet Russia … with all our power … with the object of fighting against the imperialist governments for the liberation of all oppressed peoples ...' A close examination of Kemal's letter shows that it is a manifestation of an unqualified guarantee of Turkish support to the Bolsheviks' campaign in the Caucasus. Kemal confirms: 'the Turkish government accepts the responsibility ... of compelling Georgia ... and Azerbaijan ... to enter into union with Soviet Russia, ... and [we are] ready to undertake military operations against the expansionist Armenia.'[102]

Soviet Control in Transcaucasia

While Constantinople was under British occupation and the new Turkish parliament in Ankara had already expressed its full commitment to a close co-operation with Moscow, the region of Transcaucasia became the hot spot of Bolshevik foreign policy. It was called the 'new active front' of the Soviet state to prevent the 'hostile' designs of the British in Asia Minor.[103] While the British and their allies 'concentrate all their attention upon Anatolia', the Sovietisation of the Caucasus appeared to the Bolshevik leaders in Moscow as an easier as well as timely task.[104]

In this favourable atmosphere the Bolsheviks engaged themselves in the final stages of total domination in the Caucasus with a pronounced determination. The day following the Allied occupation of Constantinople Lenin wired Smilga[105] and Ordzhonikidze, members of the Military Revolutionary Council of the Caucasian Front, and instructed them to carry out preparations for the occupation of Baku and continue to advance on Georgia.[106] Ordzhonikidze, after organising guerrilla units in Daghestan and Chechnya in the north Caucasus in the first half of 1919, had spent the rest of the year attached to the Fourteenth Red Army operating in the Ukraine. In early 1920 he had been sent back to the north Caucasus.

In the meantime the Volunteer Army, which had put the north Caucasus under its control in early 1919 had virtually collapsed in the winter of 1919–20. The Red Army units of the Caucasian front had moved in and reoccupied the north Caucasus in early 1920. Upon his arrival in the region Ordzhonikidze formalised his status as the president of the North Caucasian Revolutionary Committee, in other words the political boss of the region.

When he received Lenin's telegram in March Ordzhonikidze started to organise the offensive and first met General Tukhachevskii[107] and S. M. Kirov in Vladikavkaz on 30 March 1920.[108] General Tukhachevskii was the commander of the entire Caucasian front and Kirov was then the political commissar with the Eleventh

Army.[109] From then on the Bolshevik Caucasian offensive took off fast and encountered no major obstacles.

The creation of the high-powered Caucasian Bureau (Kavburo) in April 1920 through the direct initiative of the Central Committee of the Russian Communist Party was an unequivocal indication of the interest shown by the Bolshevik leadership.[110] With the founding of this special bureau all Caucasian affairs were put under a strictly centralised control. This move implies that Moscow believed that the time was ripe for the ultimate Sovietisation of the Caucasus. It was regarded, however, as beyond the capacity of local communist organisations. Kavburo was linked up with the south-eastern Bureau of the RCP and soon became the indisputable head of the Bolshevik campaigns in the entire Caucasus.[111]

The principal tasks of the Kavburo were identified as the establishment of Soviet rule throughout the Caucasian region, the economic unification of the area, and the administration of relations with the revolutionary movements in the Near East (particularly with the Turkish nationalists).[112] The Kavburo was attached to the staff of the Eleventh Army conducting operations in the north Caucasus. Ordzhonikidze was appointed its chairman, Kirov was the vice-chairman; the Azeri, Narimanov,[113] and the Georgian, Bdu Mdivani,[114] served as members of the bureau.[115]

Until the formation of the Kavburo, the Regional Committee of the Russian Communist Party in Tiflis had been working for about two years to establish Soviet power in the region. The function of the committee, together with the local communist parties, was now limited to organising local revolutionary campaigns.[116] The purpose was to create a facade of revolution by internal forces, not a direct occupation by the Red Army.

The first task of the Kavburo was to seize power in Azerbaijan where the communist underground had been active since the beginning of March 1920. Following clear instructions from Moscow regarding the capture of Baku events developed rapidly. The preparations by the Kavburo took place on two different levels. It encouraged open political activity and mobilisation of public opinion on one hand, and secret underground work on the other. In order to fulfil this dual task the Kavburo made extensive use of already existing networks of political commissars within the local organisations. The aim was that the advance of the Red Army in the region would be accompanied, and sometimes preceded, by the development of local revolutions organised by the Bolshevik militants. By April 1920 many local institutions, including a number of regiments of the Azerbaijani army, were put under the control of the Bolsheviks.[117] This type of work would make it easier both to prepare the local Bolsheviks to take power and to ensure that they would do it under firm control from Moscow.[118]

The leading nationalists and Turkish officers played an important role as intermediaries in manipulating Muslim public opinion in Azerbaijan. They tried to convince the Azerbaijani authorities and leading Muslim leaders that the Bolsheviks had no desire to abolish the independence of the republic and that the basic aim of their activities was to guarantee a safe passage for military aid to the national movement in Anatolia which was fighting for the independence of Turkish lands.[119] Turkish nationalists and Turkish communists organised joint meetings in Baku in April to mobilise greater public support for the Bolsheviks in Azerbaijan.[120]

On 27 April 1920 (according to Kavburo directives) the Central Committee of the Azerbaijani Communist Party, the Baku Bureau of the Regional Committee of the Russian Communist Party and the Central Workers' Conference of Baku handed the Azerbaijani government a joint ultimatum demanding its surrender within twelve hours. On the same day the news arrived in Baku that the Eleventh Army had already crossed the border the previous night.[121] The next day Ordzhonikidze and Kirov arrived in Baku followed by the troops of the Eleventh Army.[122] Power was declared to be in the hands of the Temporary Revolutionary Military Committee of Azerbaijan (Azrevkom), headed by Narimanov and consisting of Hüseynov, Musabekov, Buniatze, Alimov and Garayev.[123]

The seizure of Azerbaijan was celebrated as a 'highly important step in the development of Communism in the Near East' by Sultan Galiev[124] in the official organ of the Commissariat for Nationality Affairs (Narkomnats). 'From Azerbaijan we could hurt the British in Persia, reach out to Arabia and lead the revolutionary movement in Turkey.' These words provide a good summary of the symbolic importance of Azerbaijan, 'the window for the revolution in the Muslim East' to the Bolsheviks.[125] The importance of Azerbaijan for the Soviet state was, however, related to its strategic location and the rich mineral resources which industry in Russia depended upon. The most important of all was oil. Baku oil made Azerbaijan a strategically vital area.[126] The great field at Baku, together with the lesser but sizeable ones at Grozny in Chechnia and Maikop in Adigeyskaya, supplied 100 per cent of Russia's crude oil demand. Following the events in Azerbaijan the Narkomindel sent a statement to the Politburo claiming that 'our capture of Baku' made a shock impact upon the Western powers, who were fighting each other for the oil reserves.[127]

Following the Sovietisation of Azerbaijan, Ankara's position toward Baku became more sympathetic. Ibrahim Abilov was posted to Ankara as the representative of Soviet Azerbaijan. New consulates were opened in Kars, Trabzon and Samsun. Soon afterwards, however, the Bolshevik military campaign in the Caucasus was called off. This was related to an emergency on the western front.[128] Since February 1919 Polish and Bolshevik forces had been at war. On 25 April 1920 the chief of the Polish state, Marshal Pilsudski, had launched a major offensive against the Bolsheviks. As a result of the swift advance of the Polish army into the interior of the Ukraine, the Bolsheviks evacuated Kiev on 6 May and on the morning of the following day the Polish forces occupied the city. War with Poland tied up the Soviet armies for the next six months until 12 October when the armistice and preliminary peace treaty was signed in Riga.[129]

While the main Red Army units were engaged in the Soviet–Polish war, the situation in the Caucasus continued to be favourable for the Bolsheviks. In the summer of 1920 the remaining troops of Great Britain were evacuated from Transcaucasia and the last hope of the Allies, a proposed American mandate over Armenia, was rejected by the United States.[130] The Russian Bolsheviks were left as the unchallenged arbitrators of the fate of the region. From then on the Bolsheviks, with the support of the Turkish nationalists, determined the course of events with a Sovietisation plan across the region.[131]

In July a Turkish delegation headed by Foreign Minister Bekir Sami arrived in Moscow to give a formal status to the bilateral relations and to accelerate the shipment of arms and money to Anatolia.[132] During the negotiations the Bolsheviks drew the Turkish delegation's attention to physical arrangements. In order to facilitate further Soviet military and financial aid a land-bridge was needed between Russia and Turkey as soon as possible.[133] The material aid from Russia would be greatly facilitated by the unhampered use of the railway line between Erzurum and Baku, and the greater part of this line was controlled by the independent Armenian Republic.

The decision of the Politburo to give material assistance in terms of weapons and gold to the Turkish nationalists came on 28 July. The Politburo's statement stressed that the Soviet government was ready to help the Turks in their struggle against the Allies. The decision to prepare 'special staff … for Turkey, Iran, Afghanistan,

India and Far East' within 'the Eastern department of the Academy of General Staff' was announced in the same statement[134] which also announced the appointment of Eliava as the Soviet representative to Turkey.

While the first decisive steps had already been taken both by Ankara and Moscow towards further co-operation in the region a diplomatic development on the Allied front directly influenced the course of the Turkish–Soviet rapprochement by sharply increasing anti-British feelings in Turkey. This was the Treaty of Sèvres which was signed between the Allies and the Ottoman government in Constantinople on 10 August 1920. The main provisions of the treaty were as follows: Turkey in Europe would be delineated approximately by the Catalca lines – less than 20 miles from Constantinople; Constantinople would remain under Turkish sovereignty. But if the Turks failed to observe the provisions of the treaty, especially on the protection of minorities, the Allied forces could modify these provisions. As for the Straits, they would be open both in peace and in war to vessels of commerce, or war, and to military and commercial aircraft without distinction of flag. They would not be subjected to blockade and no belligerent right would be exercised, nor any act of hostility committed within them without a decision of the Council of the League of Nations. Finally, on the contentious Armenian and Kurdish questions, Turkey was asked to relinquish a substantial portion of eastern Anatolia to an independent Armenia and an autonomous Kurdistan.[135]

The treaty was described by Lloyd George as 'a policy of releasing all non-Turkish populations from Turkish sway'.[136] However, it must be noted that the Ottoman government, which signed the treaty, had no authority in Anatolia, and the Allies, who imposed it, were not in a position to provide troops to enforce this latter provision. It was, therefore, clear that the Treaty of Sèvres did not rest on a solid foundation; unaccompanied by real help, it was to remain an ineffectual document.

Nevertheless it made the Turkish nationalists extremely anxious about the future of the Armenian provinces in eastern Anatolia. The retaliation came two weeks later when a draft Turkish–Soviet accord was initialed in Moscow. This called for both sides to make every effort to open an unobstructed avenue between the two countries for the flow of men and material.[137] The Turkish delegation sent their observations to Mustafa Kemal on the progress made during the negotiations. The central theme of their telegram was the extent of the Soviet government's commitment to the Caucasus.

They held the view that the Bolshevik forces were ready to pounce on the Caucasian front with all their might.[138]

In September 1920, soon after the terms of the draft Turco–Soviet treaty arrived in Ankara, a full-scale war broke out between the Armenians and the Turks.[139] Since the summer of 1919 there had been a dispute between two sides over the eastern provinces of Anatolia.[140] An independent Armenia had long been seen as an important obstacle separating Kemal's army from their 'Azerbaijani brothers', and from Bolshevik aid. In January 1920 Yusuf Akçura[oglu], a well-known champion of Pan-Turkism and main ideologue behind the CUP,[141] was promoting the idea of 'the liberation of the oppressed Turks'. He was adamant that 'It is necessary to destroy Armenia which the Allies want to erect as a barrier between the two brotherly segments of Turkdom, Anatolia and Caucasus ...'[142]

According to Rawlinson, who became a prisoner in Erzurum on 16 March 1920, 'there were constant signs of preparations for a military offensive' against Armenia even before the Treaty of Sèvres. More troops were constantly arriving at the fortress of Erzurum and machine-gun practice was being carried on every day.[143] The Turkish offensive caused extreme anxiety in Armenia. On 10 September 1920 the Armenian Social Democratic Party 'GNCh.AK' sent a letter to Lenin accusing Karabekir and Kemal in person of pursuing a 'policy of genocide', similar to that of the CUP in 1915, with the aim of total extermination of the Armenian nation.[144] The letter demanded from the Soviet government the protection of the integrity of the Armenian lands from the expansionist policies of their arch enemy.[145]

There is no indication in the archives whether the Soviet government had any intention to intervene to stop the Turkish armies, but in any event no immediate protection was offered to the Armenians at this stage. The Turks moved against Armenia without any serious opposition. Kars, the citadel fought over by Turks, Russians and Persians since ancient times, was the first target. Turkish forces advanced rapidly and took over Kars on 30 October. Soon after, negotiations were opened in Alexandropol, less than 40 miles east of Kars. The rapid advance of the Turks was criticised severely in another letter to Lenin, signed this time by Eliava, Skatchko, Narimanov and Pavlovich.[146] The letter is dated October 1920 and must have been written either on 30 or 31 October since it mentions the Turkish capture of Kars. It was a protest letter, calling the Soviet government to stop its policy of alliance with the government of Kemal whose aggressive army was ordered to occupy Armenia and to exterminate the Armenian nation.[147]

While the Armenian government was discussing the stiff conditions put forward by the Turks, the Kavburo finally decided that they had to act. It seems that the Bolsheviks realised that they had suddenly been presented with an ideal opportunity to continue their campaign for a Sovietised Caucasus. On 27 November 1920 Stalin telephoned Ordzhonikidze in Baku and instructed him to commence operations against Armenia immediately.[148] On 29 November Legran,[149] who had arrived in Erevan on 11 October as the Soviet plenipotentiary, handed the Armenian government an ultimatum demanding the transfer of authority to a 'Revolutionary Committee'.[150] This was followed by the advance of the Eleventh Army which crossed the Armenian border from Azerbaijan and proceeded toward the capital.[151] Pavlovich was quick to celebrate 'the Soviet revolution in Armenia' as 'a breach in the front British capitalism is erecting against' Soviet power.[152] Faced with a double ultimatum from the Bolsheviks and the Kemalists, the Armenian Dashnak government decided to come to terms with the 'lesser evil' of Bolshevism, and on 2 December it signed an agreement with Legran, according to which Armenia was proclaimed an 'Independent Socialist Republic'.[153]

Moscow's decision to order the Red Army across the border seems to have been prompted by a determination on the Bolshevik side to prevent the complete invasion of Armenia by Turkish troops.[154] A telegram dated 24 November from the special Soviet mission in Georgia informed Moscow of the confused state of affairs and panic which resulted from the Turkish attack on Armenia. The telegram pointed out that the reluctance of the majority of the Armenian soldiers to fight accounted for the easy capture of the region of Kars and Igdir.[155] During the Turkish advance on Armenia two Turkish gunboats had been stopped in Novorossisk by the Bolsheviks and prevented from sailing southward until the declaration of Soviet Armenia. The event cast a chill over the 'mutual understanding' and caused a flow of protest statements from Ankara.[156]

The invasion of Armenia by the Red Army in order to prevent Turkish advance accounts for the first crack in the newly formed relationship between Moscow and Ankara. But both acted with reason; despite their apparent conflicting interests over Armenia they could not afford to give up 'the special relationship'. More than ever their co-operation now resembled a business partnership rather than a unity of principles. The Turks did not attempt to go further towards the historically Russian-held parts of Armenia and the Russians did not move further down into the Turkish Armenia.

Having the misfortune to lie surrounded by the Russian Bolsheviks to the north and Turkish nationalists to the south, Armenia was effectively divided into two.

After Azerbaijan and Armenia, Georgia was the third and the final domino in the Sovietisation scheme. Since the mid-1920s the local Bolsheviks, under the directives of the Kavburo, had actively fermented political unrest in Georgia.[157] On 15 December 1920 – less than two weeks after the proclamation of the Armenian Socialist republic – the Kavburo called a meeting on the situation in Georgia where it was agreed to proceed with an offensive against Georgia. After this decision Ordzhonikidze wired to Lenin requesting authorisation from the Central Committee of the RCP(B) for a Red Army invasion of Georgia. Lenin at first hesitated to approve the decision of the Kavburo, sanctioning only political aid for the Bolshevik underground in Georgia.[158]

It seems that there were primarily two reasons for Lenin's hesitation. First, the last days of 1920 and the early days of 1921 coincided with the time when Lenin was anxious to improve relations with the Western world as part of the reconstruction of the country's economy and to stabilise social life after the devastating effects of the Civil War. The establishment of regular diplomatic relations was, therefore, extremely important and the Bolsheviks were trying not to strain this delicate situation. A Soviet delegation had already been dispatched to London and negotiations were under way with the British government. An open military attack on Georgia, whose Menshevik government enjoyed some sympathy in European socialist circles, might have had a harmful effect on the prospects of the Soviet government in its relations with the West. On 21 December, at the Eighth All-Russia Congress of Soviets, Lenin urged the delegates to be 'very careful to avoid anything that may involve us in war'. When he was asked 'what is going to happen next to Georgia?' he replied with extreme caution that 'Even the Mohammedan Allah doesn't know that, but if we show self-restraint we have something to gain without running any risk.'[159]

Second, the deterioration of relations with Turkey, whose armies were anxiously waiting at the borders of Georgia, caused further suspicion about the success of the proposed attack on Georgia. Since the Turkish capture of Kars and the declaration of a Soviet Armenia Turco–Soviet relations seemed to have cooled and the future of the co-operation in Anatolia looked bleak.

Armenian affairs, however, did not cause a break in relations. There was an additional reason for increased suspicion and mistrust between the two parties. The uneasy relationship between the Kemalists and the various left-wing and pro-Soviet movements in Anatolia underwent a sharp change in the last months of 1920 resulting in a major crackdown on all potential sources of subversion. What we know about this anti-communist campaign in Anatolia comes from mainly two sources: the accounts of the contemporary press[160] and the minutes of discussions at the Turkish GNA in Ankara.[161] The information provided by various Soviet representatives in Turkey and the Caucasus confirm the extent of this campaign.[162] The campaign reached its height with the murder of 15 leading members of the Turkish Communist Party off the coast of Trabzon only one month after their arrival in Turkey.[163] These events will be dealt in some detail in Chapter 4.

The distancing of the Turkish government became evident over the issue of Georgia. Until late 1920 Ankara had been supporting the Bolsheviks' campaign to set up Moscow-controlled regimes in the region which included Georgia. This had simply been considered as a fair trade-off for weaponry and money. Mustafa Kemal had affirmed on many occasions that the existence of independent states in the region was part of a grand British design aimed at preventing further co-operation between two anti-Western powers, the Russian Bolsheviks and the Turkish nationalists. However, in November 1920 the Soviet mission in Georgia reported to Moscow that the representative of the Ankara government in Tiflis, Kazim Bey, had indicated to Menshevik Georgian government officials that Ankara was strongly in favour of keeping Georgia as an independent republic.[164] In late December Chicherin received another telegram from his mission in Georgia informing him that the Kemalists had already promised active support to the Georgian government.[165] On 18 January 1921 Chicherin informed Ordzhonikidze in Baku that the new Turkish representative in Moscow had also hinted at a Turkish intention to take an active part in Georgian affairs.[166]

The new Turkish representative in Moscow was Ali Fuat Pasha who had been active in the nationalist movement as a close associate of Mustafa Kemal since early 1919. After participating in the Amasya meeting and the Erzurum and Sivas Congresses, Ali Fuat had been made the commander of the Nationalist Armies on the Western front in September 1919. Ali Fuat's active military career came to an end in November 1920 when he was appointed the first ambassador of the Ankara government to Moscow. Ali Fuat and his large team left Ankara in early December and spent a week in Tiflis *en route* to Moscow. His memoirs reveal that in Tiflis the Georgian government asked for active Turkish support against a possible invasion of the country by the Soviet Red Army. Ali Fuat does not tell us how the Turkish delegation responded to the Georgians but he implies that Turkey could not stay aloof if Moscow decided to capture Georgia.[167]

The Bolsheviks were concerned and suspicious of the political motives and military position of the Ankara government in the early days of 1921. They were anxious to learn the real man-power and strategic plans of the Turkish army.[168] It is clear from Soviet documents that they believed that Turkish intervention might seriously endanger the Bolsheviks' campaign in the Caucasus This would have caused serious ramifications across the whole region. The Caucasian Red Army was considered 'unready for any large-scale operations' against another regular army, particularly if war in Georgia would be a prolonged one.[169]

Despite Moscow's initial reluctance, Ordzhonikidze persistently repeated in January the request of the Kavburo to advance on Georgia immediately. Kavburo's position was supported by Chicherin who presented a memorandum in which he identified the situation in Georgia as favourable for such a move.[170] Chicherin argued that unless immediate action was taken it could be too late. There was a strong possibility that the Turkish army could have advanced from the south to provide military support to the Georgian Menshevik government.[171]

Coinciding with the discussion in Kavburo Krasin reported from London that the British now saw the entire Caucasus as lying within the Soviet sphere of influence.[172] This position emerged in a special meeting between Lloyd George and Krasin on 21 December where Krasin obtained significant concessions from the British government regarding the trade negotiations.[173] This development was interpreted by the Bolsheviks as an encouragement – a free hand to the Soviet designs in the region.

This state of affairs might have forced Lenin to reconsider the situation in Georgia. Whatever influenced him to give permission to Ordzhonikidze – whether it was the possibility of a Turkish invasion of Georgia, the fear of alienating Kavburo, Chicherin's support, or the implication of Krasin's report, or perhaps a combination of these factors – once he made up his mind, he acted quickly. At the end of January instructions were issued to the Communist Party of Georgia to organise an uprising. As in the Azerbaijani and Armenian cases local communist party cells were to be used as the avant-garde of the Red Army advance. On the night of 11 to 12 February, at the instigation of Ordzhonikidze, an uprising broke out at Borchalo against the Menshevik Georgian government. Two days later the Central Committee dispatched to the Revolutionary-Military Committee of the Eleventh Army and to Ordzhonikidze the permission to come to the assistance of the rebels.[174] The Eleventh Army crossed the Georgian frontier on 16 February and on 25 February entered Tiflis, the ancient capital of Georgia.[175]

Similar to the pattern of the Soviet intervention following the Turkish offensive against Armenia three months earlier, the Turkish nationalist army intervened in Georgia following the advance of the Red Army. On 5 March, eight months after the British had withdrawn their troops,[176] the Turkish armies entered Batum and on 8 March 1921 the Turkish government proclaimed the annexation of Batum.[177]

Meanwhile, Chicherin on 1 March, 1921, wrote to Krestinsky that 'relations with Turkey have reached an extremely critical point ... The results of such a sharp change might cause significant consequences for our eastern policies.'[178] The next day he officially protested the Turkish occupation of Georgian territory in his note to Ali Fuat Pasha, Turkish representative in Moscow.[179] In response Ali Fuat handed two notes to the Soviet government. The Turkish notes claimed that the Turkish armies were merely providing security to the Turkish minority in Batum and strongly criticised the Soviet military operations in the region. It was declared that the Soviet military operations were creating undesirable circumstances which would alienate the Muslim population of the Caucasus.[180] In his memoirs Ali Fuat writes that the Turkish forces moved into Batum in response to an official request by the Georgian government.[181]

The Georgian government, like the Armenian, was under fire from both sides and forced to make a choice. On 18 March a cease-fire agreement was signed between the Georgian government and the Georgian Revolutionary Military Committee (Revkom), which had been set up by the Communist Party of Georgia on 14 February 1921.[182] The next day, in accordance to the cease-fire agreement, the Caucasian 'Iron' Cavalry Division commanded by D. P. Zholba advanced on Batum, and together with the Georgians they threw back the Turkish troops to their former positions.[183]

With the seizure of Georgia, the process of Sovietisation of the separated borderlands in the Caucasus was completed by the Bolsheviks. By mid-1921 the region was no longer a power vacuum. A regional equilibrium was finally reached. With the establishment of Soviet power in all three Transcaucasian states, the region ceased to be a crisis area, at least on the international agenda.[184] The British decision to withdraw troops in early 1919 had eliminated the possibility of a Soviet–British clash in the Caucasus and facilitated a transition from acute military rivalry to a search for a less dangerous and more viable political relationship. Once the tensions relaxed, the question was whether or not the new situation signalled the beginning of an official understanding to regularise the continuing rivalry between Britain and Soviet Russia.

4

Agreements on Two Fronts: 16 March 1921

> Enver wishes to publish in Moscow two Turkish papers. One of them aims at the Turkish people, and the other at the other Muslim countries ... The papers will promote the struggle for emancipation of the East and will emphasise on the significance of Soviet Russia for the liberation of all oppressed peoples. The [forthcoming] agreement with England prohibits all propaganda activities outside the boundaries of Russia, but it does not say anything about such activities within Russia. The delivery of the papers to the Muslim world will not be done by us, so [the publication and the delivery of the papers] can not be seen as a breach of our agreement [with Britain] ... This extraordinary politician [Enver] understands the situation well and he knows that we require him. We propose to allow him and to render him all necessary assistance to publish the papers [in Moscow] ... [Chicherin, 18 April 1921][1]

The Anglo–Soviet Trade Agreement

March 1921 may well be regarded as one of the historic months in the history of British–Soviet–Turkish relations. This is not only because in this month the Caucasus ceased to be a crisis area in the relations between these countries but also because it witnessed the first official agreements of the Soviet government with London and Ankara: the Anglo–Soviet Trade Agreement[2] and the Soviet–Turkish Treaty of Friendship, both signed on the same day, 16 March 1921.

The Anglo–Soviet Trade Agreement was signed in London on 16 March 1921, less than a month after the Red Army entered Tiflis. It was the culmination of a long process of negotiations which went back to the 25 March 1920 when a Soviet Trade Delegation had been formed to conduct preliminary negotiations concerning trade and peace with the West.[3] The head of the delegation was Leonid Borisovich Krasin, formerly an engineer and manager in the electro–

chemical industry.[4] Before the Bolshevik Revolution he had held high-level managerial posts both in Russia and abroad, including managing director of the Russian subsidiaries of the German electrical complex of Siemens Shückert.[5] He had also worked in Baku with the joint-stock company 'Elektrosida' as assistant director in charge of the construction of a large electrical power station on the shore of the Caspian Sea. He had originally become interested in Marxism largely because of his desire to play a role in the modernisation of Russia. Krasin was actively involved in the Russian Social Democratic movement and had led Marxist propaganda in workers circles since 1889. He was described by Trotsky as 'old revolutionary, eminent engineer, brilliant Soviet diplomat, and above all intelligent human being'.[6] Krasin was appointed to lead the Soviet delegation because of his extensive contacts with West European business and finance circles, his familiarity with European countries, and his solid reputation as one of Soviet Union's foremost engineers and industrial managers.

Krasin arrived in London on 27 May 1920. The next day the Conference of Ministers met under Lloyd George's chairmanship to discuss the imminent negotiations with the Bolsheviks. Sir Robert Horne, who had been appointed president of the Board of Trade just two months previously, and Sir Laming Worthington-Evans, minister without portfolio, were put in charge of the negotiations.[7] At the Conference Curzon asked that they should not restrict their discussions to trade. He insisted that the opportunity of Krasin's visit should be seized in order to come to a political understanding on the many points with which the British government were at issue with the Soviet government in different parts of Europe and the East. Curzon's main concern was related to the threat stemming from the Soviets' increasing influence in the East. Only a month had passed since the Red Army's entrance into Azerbaijan and there was disturbing news every day regarding the activities of the Bolshevik agents in the other parts of the Caucasus, Turkey, Central Asia and (most important of all) India. The Conference of Ministers agreed to Lord Curzon's suggestion that the negotiations should aim for a comprehensive political agreement with the Soviet government.[8]

The primary phase of negotiations started with the first meeting of the members of the Soviet delegation with Prime Minister Lloyd George at 10 Downing Street on 31 May 1920. In this meeting Lloyd George was accompanied by Bonar Law, Lord Curzon, Sir Robert Horne, and Mr Harmsworth. The negotiations continued throughout June.[9] On 30 June Lloyd George handed the Soviet delegate a written statement spelling out British preconditions for further discussions.[10] After this Krasin returned to Moscow to discuss the British terms. On 7 July the Politburo met and agreed to accept the British demands. The same day Chicherin notified Lord Curzon about the decision. Two days later, on 9 July, Chicherin wrote to Curzon again stating that Moscow was willing to consider the agreement as an armistice between Russia and Great Britain, which he hoped would soon lead to a conclusive peace.[11] Thus the first round of the Anglo–Soviet trade negotiations ended with a strong sense of optimism.

Before the second phase of the negotiations, however, an important problem emerged which put them in serious danger. This was related to a sudden Soviet advance in the West into Poland.[12] On 7 July Poland appealed to the Allied governments for international aid against the Soviet aggression.[13] The Red Army, less than a month after abandoning Kiev to the Poles in early May, had made a sudden and unexpected break-through in June and forced the Poles to begin evacuating Kiev. On 4 July another Bolshevik offensive was launched across Berezina, 350 miles east of Warsaw. On 13 June the Polish army was in full retreat. At the beginning of July the Red Army launched a powerful offensive on the western front which brought the Bolsheviks to the gates of Warsaw, creating panic in the West.[14]

On 11 July Lord Curzon dispatched a note to Moscow demanding a reply within one week. Curzon's letter demanded an immediate armistice between Russia and Poland. He suggested that Polish troops would be withdrawn to the ethnic border (a Polish–Russian frontier running roughly from Lwow straight north to the Lithuanian border) if Soviet soldiers would stop 50 kilometres east of it. Curzon further proposed a conference in London, under the sponsorship of the Peace Conference, which would work out the details of a general peace between Soviet Russia and its neighbours – excluding Romania. Curzon's note agreed to the continuation of trade negotiations as soon as the Soviet delegation returned.[15]

In this new and apparently more complicated situation Moscow decided to send a second delegate who would deal with diplomatic questions and without whose consent Krasin would not have the right to make any political announcements.[16] It was decided that Lev Borisovich Kamenev, Miliutin and Theodore Rothstein would accompany Krasin. Kamenev[17] was a senior member of the Politburo, chairman of the Moscow Soviet and one of Lenin's closest friends; Miliutin was vice-chairman of the Supreme Soviet Council of National Economy; and Rothstein[18] was the Bolsheviks' main undercover political agent in Britain after the expulsion of Litvinov in 1918.[19]

Chicherin, in his reply to Curzon's note, reiterated Russia's readiness to enter into an armistice leading to a permanent peace, but he pointed out that the Polish government should speak for itself and approach Moscow independently and directly.[20] Chicherin also added that 'in order to reach a complete agreement' the Soviet government was ready to send its enlarged delegation to London. When, on 17–18 July, the Soviet reply was received in London it was considered as a measure of Moscow's moderation. After reading Chicherin's reply, Lloyd George was reported to comment as 'I don't call that unreasonable ...'[21]

However, the newly enlarged Soviet delegation was not able to proceed to London at once. On 20 July Lord Curzon informed Chicherin that under the circumstances, with the Red Army still advancing toward Warsaw, they could not expect any progress in the trade talks. He confirmed that the visit of the Russian delegation should be put back until an armistice agreement was achieved between Russia and Poland.[22] On 23 July Chicherin replied to Curzon saying that an armistice between Poland and Russia was a newly improvised precondition which Britain had added to the previous four accepted in the first week of July. Nevertheless, he assured the British Foreign Office that such a precondition would be met soon as the Poles had already indicated that they would agree Moscow's request for an armistice.[23] Indeed, the next day two telegrams were dispatched from Warsaw informing the Kremlin that the government of Poland was ready to discuss an immediate armistice and the opening of peace negotiations.[24] This was followed by Lord Curzon's announcement on 26 July that a warship would now escort the Soviet delegation from Reval in view of the Soviet acceptance of the Polish proposal.[25] In the end, after about a month's delay due to the Soviet–Polish war, the Soviet delegation met Lloyd George again on 2 August 1920. The negotiations continued throughout August. But this time a scandal prevented them from going any further.

Kamenev and Krasin had brought with them to London precious stones, mostly diamonds and platinum, worth around £40,000. In London they sold most of these stones to finance a number of pro-Soviet bodies, including the *Daily Herald*, a newspaper which had existed since 1908 and was close to the Labour Party. On 20 August Kamenev informed Chicherin that a large sum of money (£75,000) had been given to the *Daily Herald*. Smaller sums were given to other left-wing periodicals and to the Communist Party of Great

Britain on its foundation in August.[26] Most of the Soviet communications – coded telegrams and wireless messages between the delegation and the Soviet government – including those about financial support to the *Daily Herald*, were intercepted by British intelligence. The British government decided to publish a selection of the intercepted messages about the *Daily Herald* and sent eight of them to all national newspapers except the *Daily Herald*.[27]

A number of ministers, including Curzon and Churchill, wanted to break off negotiations with the Soviet delegation. In order to save the negotiations, Kamenev decided to return to Moscow, and left Britain on 11 September. On 14 September the British Cabinet decided to instruct Wise to complete the draft Russian Trade agreement.[28] On 20 September the Central Committee of the RCP(B) authorised Krasin to negotiate the conclusion of the treaty.[29] The draft trade agreement prepared by the Interdepartmental Russian Trade Committee was formally handed to Krasin on 29 November.[30] On 11 January 1921 Krasin left London again for Moscow, taking with him a draft of the proposed trading agreement. The Soviet government replied on 5 February 1921 accepting the proposed agreement in its basics. Krasin returned from Moscow on 4 March, 1921, and the Trade Agreement was duly signed on 16 March 1921.[31]

The normalisation of relations between the Soviet state and the outside world was the primary motive behind the Soviets' eagerness for a trading agreement. World War, Revolution and the Civil War had left Russia an economic ruin. The revolution had triumphed, but at a great cost. The move to peace in the ruined country proved difficult. It was time to redeem the promises made to the population. The end of the Civil War did not end the Russian people's suffering and the grave problems which the Bolshevik government faced. In Lenin's words, the country's economy was 'ten times or more worse off than before the war'.[32] Agricultural output was also severely reduced. Large sections of the countryside were in revolt, the mass of the peasants were in a state of utter exhaustion, proper relations between the town and the country were disrupted, and the peasants were refusing to deliver their product to the cities. During the war and the Civil War at least 7 million people had died from hunger and disease. In the first year of the peace, 1921, one of the worst famines in Russia's history claimed 5 million more victims.

To Lenin the opening of normal trade relations with the Western world implied a decisive turning point. 'Up to now, we have been adapting ourselves to the tasks of war; we must now adapt ourselves to the conditions of peace.'[33] It was hoped that a thriving foreign trade, and even foreign loans, would materialise which would ease the pressure upon the economy. 'To sign a trade pact' would provide 'some of the essentials for the restorations of the railways (i.e., locomotives), for the rehabilitation of industry, and for electrification.'[34] E. H. Carr writes that of 50,000 miles of railway in European Russia less than a fifth remained undamaged by the war or civil war, and more than half the country's 16,000 locomotives were out of order.[35]

Russia's economy had to be revived. This was the central theme of the Tenth Party Congress, which met on 2 March 1921, only two weeks before the signing of the trade agreement with Britain. The Tenth Party Congress is noteworthy chiefly for its resolutions on the introduction of the New Economic Policy. With the New Economic Policy, the Bolsheviks abandoned an economic policy based on centralised control and force. The New Economic Policy brought back the marketplace and traditional market initiatives. The peasants were free to consume or sell their product on the open market and a series of legal and political reforms were introduced to attract Western businessmen and to facilitate economic recovery. The New Economic Policy gave rise to hopes of new prospects. The shift in economic policy made the Bolshevik regime all the more anxious for a trade agreement with Great Britain.

The Anglo–Soviet Trade Agreement not only implied the prospect of economic benefits but also, in White's words, 'represented the first resumption of relations *de facto* if not *de jure* between Soviet Russia and a major capitalist power'.[36] The 16 March 1921 signalled the end of outright antagonism and marked a formal acceptance that Bolshevik control over the former Russian Empire was almost complete and that the international activity of the new Soviet state, even in those territories 'liberated' from Russian control after 1917, would have to be accommodated within the ordinary conventions of great power diplomacy.[37] Lenin considered commercial relations with the capitalist powers to be a necessary foundation for the security of the Soviet regime. It seemed logical to the Soviet leaders that it would be more difficult for capitalist states to go to war with Soviet Russia once they had entered economic relations.[38] In this fashion, the Anglo–Soviet Trade Agreement functioned as the cornerstone of what came to be known as the New Economic Policy in Soviet foreign policy: peaceful coexistence with the capitalist states as the basis of Soviet security.

On the British side both economic and political motives played a major role. Problems of domestic and international reconstruction brought into sharp focus the relations with the Soviet government. Growing social and economic difficulties, war weariness, political opposition – as well as the many frustrating and inconclusive aspects of the Russian problem – contributed to the conclusion of the agreement with the Bolsheviks. Lloyd George aimed at aiding Britain's ailing export industries and stabilising his position at home by producing a trade agreement with Moscow.

Although it was clearly impossible wholly to separate economic from political considerations, the wider concern of the British about the Soviet regime was related to its destabilising impact. The main fear was of propaganda. If it were not feasible militarily to defend the Eastern Empire against the Bolshevik threat, it might prove possible to limit the danger by requiring from the Bolsheviks – in return for an agreement to enter into trading relations – an undertaking to refrain from agitation and propaganda hostile to British interests. In this way the British sought to marginalise Soviet influence in Asia. During the first meeting with the Soviet delegation the British prime minister referred 'to the action which had been taken by the Bolsheviks in supporting armies and states who were hostile to British interests.' Lloyd George further specified that 'the Soviets were encouraging Turkish nationalists to make trouble … in Asia Minor.'[39]

This issue attracted a good deal of attention in the course of the trade negotiations. We learn from a note appended to a confidential letter from the director of British Intelligence to Mr Heathcote-Smith of the Foreign Office that the Bolshevik mission in London was directly asked about Moscow's relations with Kemal.[40] The first and most important of the conditions which had been specified by the British government in its note of 30 June 1920 was that each side should refrain from taking part in any hostile actions or measures directed against the other and from conducting any kind of official propaganda, direct or indirect, against the institutions of the other side. The Soviet government should in particular refrain from any kind of attempt, by means of military action or propaganda, to urge the peoples of Asia to any kind of hostile actions directed against British interests or against the British Empire.[41]

The Bolshevik leaders were well aware of the sensitivity of this issue. Lenin clearly stated that 'She [Britain] is a colonial power, with too great a stake in Asian politics, and is sometimes too sensitive to the success of the Soviet power in certain countries lying near her colonies.'[42] In early June Trotsky, in a letter to Chicherin, was emphasising that 'a potential Soviet revolution in the East is now advantageous for us chiefly as a major item of diplomatic barter with England'.[43]

The Congress of the Peoples of the East, which assembled at Baku in September 1920, gave much concern to the British. The summons to the congress came from the Comintern and was first published in *Izvestiia* on 3 July. It was addressed exclusively to the 'enslaved peoples of Persia, Armenia, and Turkey'. Turkish people were called to resist the Allied powers who were controlling Constantinople and Western Anatolia.

> Peasants of Anatolia! The English, Italian and French governments have kept Constantinople under the fire of their guns; they have imprisoned the Sultan, have forced him to agree to the dismemberment of purely Turkish territory, and have handed over Turkish finances to foreign financiers, in order to facilitate the plundering of the Turkish people impoverished by six years of war. Peasants of Anatolia! You are urgently called to the colours under Kemal-Pasha, in order to fight the foreign invasion, but at the same time we know that you are trying to form your own national party, your own peasants' party, which would be able to continue the fight in the event that the Pashas should continue peace with the rapacious Entente.[44]

The idea of convening a Muslim congress originated with A. Zeki Velidi Togan,[45] a prominent Bashkir from Muslim Central Asia. Togan was in Moscow in 1919 and worked closely with Karl Radek before moving to Central Asia in spring 1920. To Radek Moscow was the obvious choice for such a congress to bring the leaders of the Muslim East around the Bolshevik cause. Togan was in favour of choosing a place in the East – possibly in Central Asia. Baku was chosen as the place for the congress mainly because it was already serving as a refuge for many communists of Armenian, Georgian, Persian, Turkish and other nationalities of the East. Because of its geographical location it was readily accessible to the Muslim peoples of Soviet Russia and the Iranians, as well as the Turks and the Arabs from the south.

Ordzhonikidze and Zinoviev were chosen by Lenin as the main organisers of the congress and those Bolsheviks who had some experience of Eastern affairs were selected to work with them. Thus Anastas Mikoyan, Nariman Narimanov, and Mir Said Sultan Galiev made up the 'Orgburo' which fixed the rules of participation for the delegates.[46]

Not all the communists were in favour of convening a Muslim congress with the aim of increasing the prestige of the Bolshevism among the peoples of the East. Prominent Bengali communist M. N. Roy in his memoirs published in India after his death, made it clear that he opposed the idea of the Baku Congress. Roy argued with the Bolshevik leaders that it could only serve as a means of agitation which alone was not enough to bring about a revolution in the East. He recalls in his memoirs:

> Lenin smiled indulgently on my cussedness; Zinoviev was angry at the audacity of the upstart crossing his will; Radek ridiculed my precocious seriousness. It might not yield any lasting results, but why forgo the fun of a picturesque show which was sure to give the then British Foreign Secretary, Lord Curzon, some sleepless nights.[47]

The high command of the Bolshevik leadership must have shared Radek's humorous view and the official consent was given to the congress, which opened on 1 September 1920.[48] In all 1,891 delegates attended,[49] of whom 1,273 were said to be communists, 226 non-party, and only 55 were women. The Turks were the largest national group. In addition to 235 Turkish, 192, Persian, 157 Armenian, and 100 Georgian delegates specifically summoned by the Comintern – there were also 8 Chinese, 8 Kurds and 3 Arabs.[50]

The essential aim of the congress was proclaimed to be to initiate an anti-imperialist platform among the Eastern nationalities.[51] The importance of support to national liberation movements like Mustafa Kemal's was reiterated many times from the platform. It is interesting that the spirit of united front against 'the foreign imperialist yoke' was further strengthened by a declaration from the isolated figure of Enver Pasha, and an enthusiastic speech given by the official representative of the Ankara government.[52]

The Baku Congress approved in principle the issuing of an 'Appeal to the Peoples of the East'. This was a document obviously drafted for use as an instrument of propaganda throughout the Muslim world. It was directed entirely against Britain, the power regarded by the Comintern as the one great empire which had emerged from World War I with the strength and intention to dominate the 'oppressed' peoples of the East.[53]

Anti-British agitation rose to its highest level with the last meeting of the Baku Congress of the Eastern Peoples when the 26 Baku commissars were praised as the victims of British imperialism. A funeral ceremony was organised that day with the participation of all the delegates together with the families of those 26 Baku commissars who had been killed two years previously. Party and state figures from Azerbaijan, delegates to the congress and representatives of the Comintern delivered anti-British speeches for the memory of the commissars.[54]

When the Baku Congress ended on 8 September it was decided that it would meet in the following years and a permanent centre would be established for the organisation of communist activities in the East. Neither of these decisions were, however, fulfilled.[55] Carr claims that the reason for this failure may have been related to the commitments the Soviet government undertook with the signing of the Anglo–Soviet Trade Agreement in March 1921.[56] If one may take Carr's analysis further it might not be an exaggeration to say that the whole episode of the Baku Congress may well have been designed to put pressure on the British government to sign a trade agreement with Moscow in the first place.

The proceedings of the Congress were watched anxiously by the British through their contacts among the delegates. As anticipated by Radek, it gave Curzon some sleepless nights.[57] With the first commercial agreement with the Soviet government on 16 March 1921 the British government hoped that anti-British propaganda could be checked. There was no reason to believe that the agreement would end the Bolshevik propaganda immediately. The wider expectation was that with time the Bolshevik regime would mellow and become easier to co-exist with.

According to the agreement, both sides undertook to refrain from hostile acts and propaganda against each other.[58] In the final agreement it was particularly emphasised that the Soviet government would stop all form of action and propaganda to encourage the peoples of Asia to take any form of hostile action against the British Empire.[59] The rest of the text was aimed at the technicalities of a new trading agreement with the debt question being left for a formal peace treaty to be signed later.[60]

The Soviet–Turkish Treaty of Friendship

On the same day as the Anglo–Soviet Trade Agreement was signed, the Turkish nationalist government and the Soviet Russian government signed an agreement in Moscow. According to Yusuf Kemal, the head of the Turkish delegation, the treaty was signed either on 22 March or on 23 March 1921. In his memoirs Yusuf Kemal states that he had suggested to Chicherin that the Turco–Soviet friendship treaty should carry the date of 16 March as the date of its signature to mark the opening of a new era of friendship between the two countries on the first anniversary of the formal occupation of

Constantinople by the Allied armies on 16 March 1920. Yusuf Kemal says that this proposal was accepted by Chicherin.[61] It is also a strong possibility that the date of the agreement might have been accepted as 16 March, the same day as the Anglo–Soviet Trade Agreement, in order to clear the suspicions of the Turkish side about a British–Soviet rapprochement behind their back.

The Anglo–Soviet negotiations had indeed caused deep anxiety in Ankara whose agents both in European capitals and in Moscow had been ordered to follow the course of the negotiations in detail.[62] In late 1920, just after the draft treaty agreement was formally handed to Krasin by the British, Ahmet Muhtar, the Turkish foreign minister, sent Chicherin a telegram, stating his concerns and asking information about the details of the negotiations.[63] Chicherin immediately replied Ahmet Muhtar by saying that 'Britain wants us ... to cease all our contacts with Turkey', but he affirmed that the Russian Soviet government would not yield to this pressure.[64] He reiterated the view of Soviet Russia being Turkey's saviour just before the signing of the Anglo–Soviet Trade Agreement.[65] By accepting the readjustment of the date of the Turco–Soviet agreement as 16 March, on the same day with the Anglo–Soviet Trade Agreement, Moscow might have aimed to regain the trust of the Turkish side.

The Turco–Soviet 'Treaty of Friendship and Brotherhood', as its own preamble describes it, was the product of more than eight months' negotiations. Following the first official letter of the Ankara government to the Soviet government in April 1920, the nationalists decided to send a delegation to Moscow to establish an official understanding between two governments. It was hoped that an official agreement with the Soviet government would provide Ankara much needed help and financial support. Bekir Sami,[66] who had just been elected foreign minister by the Grand National Assembly in May 1920 was made the head of the first Turkish delegation to Moscow.[67]

The delegation arrived in Moscow in the second half of July, soon after the first round of the Anglo–Soviet trade negotiations ended with an initial agreement in London. The Turco–Soviet negotiations commenced on 24 July and lasted one month- at which time a draft treaty of friendship was initialed by Bekir Sami and Chicherin. In this first round of the negotiations it soon became apparent that the major differences between the sides were related to some of the Armenian provinces that the Turkish side claimed as theirs. The Soviet delegation demanded the transfer of Van and Bitlis *vilayets* to Armenia.[68]

Although the negotiations did not go further regarding the Armenian provinces in eastern Anatolia, the first eight articles of the proposed treaty were prepared and accepted both by the Turkish and the Soviet representatives on 24 August. According to this preliminary agreement the Soviet government consented not to recognise as valid any agreement not approved by the Ankara government. It was also agreed to consider all previous agreements between Russia and the Ottoman Empire null and void. In order to secure the opening of the Straits and freedom of passage through the Straits for the commerce of all nations, the parties further agreed to entrust the final elaboration of an international statute for the Black Sea and the Straits to a special conference of delegates of the littoral countries.[69]

Following the signing of the draft agreement, the military experts of both sides met on 27 August regarding the details of the Soviet military help to Ankara. At this meeting it was decided that for the time being military help would be sent through the Black Sea, since the land route that went through the region of the Armenian republic had not been yet controlled by the Bolsheviks.[70] This was a sound promise but not enough for the Ankara government which was faced with all sorts of financial and military difficulties as a result of powerful campaigns of the Greek Army in Anatolia. Since the Black Sea was still under the control of the British, whose headquarters were in Constantinople, it was apparent that the military help from Moscow would not be satisfactory for the Turkish nationalists.

When the initial expectation of Ankara for a substantial shipment of military material from Moscow was not fulfilled the general attitude of the Turkish nationalists started to change. The Bolsheviks were blamed for not honouring their promise to Ankara. Mustafa Kemal in his speech at the Grand National Assembly in late August 1920 informed the nationalist deputies that the Bolsheviks were extremely reluctant to enter into a full commitment in their relations with Turkey.[71] This critical atmosphere was strengthened when the news arrived in Ankara about Moscow's demand to return Van and Bitlis *vilayets* back to the Armenians.[72]

There was a further and more alarming reason which reinforced Ankara's suspicion about the intentions of the Soviet government in this period. It was related to the increased activities of local communists and pro-Soviet propaganda in Anatolia. Mainly due to the initial promising atmosphere of friendship between the Bolsheviks and Turkish national movement, a number of left-wing movements had gained momentum in Anatolia in the spring and summer months of 1920. Among these the Green Army Association (Yesil Ordu Cemiyeti) was the most prominent radical organisation.

The association was founded in Anatolia in the spring of 1920 'to liberate Asia from the penetration and occupation of European imperialism'.[73] The founders of this organisation were initially on friendly terms with Mustafa Kemal. However, when Ethem the Circassian joined the organisation it was regarded as a possible threat to Kemal's authority. Ethem was one of the most prestigious leaders of the early period of the Turkish national resistance and his powerful rural guerrilla movement, Kuva-yi Seyyare, undertook most of the active resistance in Anatolia up until late 1920. The political wing of the Green Army Association set up a group among the deputies of the Grand National Assembly called the People's Group (Halk Zümresi). The prospect of a strong political organisation in the Grand National Assembly prompted a hostile attitude from Mustafa Kemal, and it was not long before Kemal accused the group of being a dangerous enterprise.[74]

There was another organisation founded in the summer of 1920 in Ankara named the Illegal Turkish Communist Party. This was a strongly pro-Bolshevik organisation and some Turkish communists recently returned from Russia took an active part within its ranks. According to some accounts Sherif Manatov,[75] the Soviet representative in Anatolia, played a major role in the formation of this party.[76] In a couple of months of its existence the party began to search for new outlets in order to broaden its base and to acquire a powerful position. As a first move it became an officially recognised legal party, renaming itself People's Communist Party (THIF) on 7 December 1920.[77]

However, by far the most well-prepared and well-connected communist movement was the Moscow-supported Communist Party of Turkey (TKP) organised and led by Mustafa Subhi since 1918. After the overthrow of independent Azerbaijan Mustafa Subhi and a number of close colleagues who had arrived Odessa from the Crimea in May 1919 went to Baku on 27 May 1920. Subhi organised a communist group among the Turks in Baku, and proceeded to open new branches of the party in other places – including Constantinople, north Anatolia, Nakhichevan and north Caucasus. Furthermore Subhi's group participated in the Baku Congress of September 1920, promoting the Communist Party of Turkey as a significant force in the anti-imperialist fight in Anatolia.

Immediately after the Baku Congress, on 10 September 1920, the

First General Congress of Turkish communists met in Baku under the leadership of Subhi.[78] Seventy-four delegates participated in the Congress, 51 of whom were identified in the documents of the Congress as from Constantinople and Anatolia. The foremost aim of the Congress was to unite various Turkish communist groups under the Subhi's Communist Party of Turkey.[79] This congress has been recognised by the official histories of the Communist Party of Turkey as its founding conference. Given the increasing pressure of the Allied occupation in Anatolia and the impressive rise in the self-confidence and determination of Subhi's party, it was not surprising that the party approved to shift the centre of its activities to Anatolia.[80]

The Baku Congress and the increased activities of Subhi's group towards Anatolia became a source of serious concern in Ankara. The slow progress of the negotiations in Moscow was interpreted in the light of these increasing activities of the communists in Turkey. Mustafa Kemal, in a letter to Ali Fuat on 14 September 1920, expressed his bitter feelings towards the Bolsheviks for organising the Communist Party of Turkey and aiming 'to make a social revolution' in Turkey. In the same letter Kemal added that if the Soviets saw any prospect of communist success in Turkey, they would not initiate material aid to the Ankara government. What if the Soviets declared they would maintain contacts with the Ankara government through the Communist Party of Turkey? Kemal remarked that 'communist organisations in Turkey were therefore completely against the interests' of the Turkish national movement and had therefore to be 'brought to a halt and kept at a distance whatever the cost'.[81] Indeed it was Kemal's firm belief that there had never been any potential for a Bolshevik-type revolution in Turkey, and never would be in future. Yet the problem at that particular time, when Ankara was expecting essential material support from Moscow, was to demonstrate to the Soviet government that this belief was well-founded without damaging the close relationship and putting Soviet help in jeopardy.

Since the official Turkish documents of this period are still closed and in the personal accounts of the leading personalities of the Turkish national movement the question of Turco–Soviet relations is treated as hardly more than a footnote, there is very little information regarding the inner discussions of the Turkish nationalists. What is known is that at this point Mustafa Kemal plotted a shrewd and cunning venture. On 18 October 1920 he instructed some close associates to set up an official communist party (TKF) in Ankara and through this party applied to join the Comintern. Although the Comintern refused this application, it was obviously an act geared to secure Bolshevik help – albeit in an unconventional way with a puppet party.[82] The official party identified Bolshevism as a completely Russian experience, therefore it did not claim to be Bolshevik. On the other hand, although it defined itself as communist, it was a very peculiar type of communism aiming for a 'unique Anatolian regime' which was not to be based on class struggle but would organise rich and poor together to fight the foreign oppressors.[83] It was of a typical populist jargon under the tag of 'communism' for the sake of its possible emotional appeal to the Soviet state.

On 31 October in a telegram addressed to Ali Fuat, Kemal repeated his belief that communism was not a feasible option for Turkey. He then went on to explain why he proceeded to set up an official communist party:

> … that this movement[communism] was being diffused in our country from internal and external sources and aiming at various goals, and unless necessary measures were taken, the peace and unity of the Turkish people would be put in jeopardy. Thus it was concluded that the wisest step would be to get some reasonable friends to form a communist party under the guidance of the government.[84]

It is obvious from the above statement that an official party was a convenient way to keep any dubious political activity under control and at the same time capitalise on any potential public sympathy towards the Soviet state. By taking this measure Kemal at first achieved a significant degree of success. The Green Army Association disbanded itself and Ethem the Circassian was persuaded to join Kemal's official party. His paper *Yeni Dünya* (New World) became the official organ of Kemal's official communist party. However this agreement did not last long.[85]

Towards the end of the year, mainly as a reaction to the efforts of the Ankara government to reorganise the rural guerrilla forces loyal to Ethem under a regular army, Ethem rose up against Kemal.[86] Meanwhile *Yeni Dünya*, now the official paper of Kemal's party, began to openly criticise the campaign of the government against Ethem and railway workers were called out on a strike to stop the transport of troops sent to quell Ethem's rebel forces. In late 1920 and early 1921 the owner of the paper and his close associates, as well as many other socialists, who were charged with being connected to Ethem, were duly arrested.[87] At the same time the leader of the Moscow-supported Communist Party, Mustafa Subhi, and some other leading figures of the party, left Baku and set out for Ankara to implement their decision to transfer the activities of the party to Anatolia. It was sheer bad timing. The group entered Turkey in late December, but could not proceed further than Trabzon on their pre-determined trip to Ankara. On 28 January Mustafa Subhi, his wife, and 14 other leading communists were put in a boat and sent back to Batum by sea *en route* to Baku. Immediately after they embarked another boat left the harbour and overtook the first one. All that is known of the events that followed is that no one[88] on the first boat survived.[89]

According to a Comintern investigation report a certain Osman Agha, from the Giresun area, was personally responsible for the murder of the Turkish communists, what is known as the 'Black Sea incident'. He was one of the influential brigand leaders in the Black Sea region and loyal to Mustafa Kemal. Osman Agha and his men were said to have conducted attacks against the Armenian and Greek villages in the eastern Black Sea coast under the directives of the Ankara government.[90]

The Ankara government categorically denied any involvement in this brutal plot and insisted that this was a maritime accident. According to Tunçay some existing telegrams (found in the Archives of the Institute of the History of Turkish Revolution) show that there was a direct link between the plotters of the murder, the Kemalist governor of Erzurum, and Kazim Karabekir.[91] In the early days of January 1921, whilst the group was still in Kars, Karabekir sent a number of telegrams to Hamit bey (the governor of Erzurum) and demanded the group to be directed to Trabzon after Erzurum for deportation. Hamit bey, on 16 January, informed Mustafa Kemal about this decision. Kemal's confirmation was received on 18 January.[92]

The available documents confirm beyond doubt that the Ankara government had a substantial role in the 'Black Sea incident'. Karabekir and Hamit Bey – one the most prominent nationalist commander, the other a very important local representative of the Ankara government – conspired not to let Mustafa Subhi into Anatolia. Mustafa Subhi and his group were directed to Trabzon where they were put on a boat for Batum.[93] The telegrams confirm that Mustafa Kemal himself was fully aware of this arrangement. However, the respective roles of Mustafa Kemal, Karabekir and Hamit bey in bringing about the fatal end of the Turkish communists is not clear. There exist no documents or any conclusive evidence as to who master-minded the final touch. Whether murder was inclusive in the plan drafted by Karabekir and Hamit bey remains a mystery. The actual nature of the arrangement confirmed by Mustafa Kemal is also obscure and open to speculation.

When the news arrived in Moscow the RCP(B) forwarded an official statement to the party members only. Basically it was an information statement explaining how the Turkish communists were killed. The central theme of the statement, however, was the dangers of left-wing, arbitrary, and adventurist initiatives.[94] Although the statement did not connect this accusation with the case directly the implication was clear. Moscow did not share the optimism and the decision of the Turkish communists to go ahead on their own and move the party activities to Anatolia.

The 'Black Sea incident' did not make a serious impact upon Turco–Soviet relations. It was noted and put aside by both sides in an almost statesmanlike fashion. The experience is, however, significant and rich in lessons. The murder of the Turkish communists in the early days of 1921 represents the first example of the failure of a peculiar Soviet dilemma in the East – to support the anti-communist leadership of national liberation movements and at the same time to sponsor and organise the local communist movements to overthrow them via the Comintern. When the Kemalist leadership openly started to root out all communist activities in Turkey the Soviet government chose to continue its official policy of co-operation with Ankara regardless of the fate of the local communists loyal to Moscow.[95]

Indeed at the same time as the conspiracy against the Turkish communists was thickening, Moscow was expecting to meet a new Turkish delegation for the continuation of the discussions that had come to a deadlock in September 1920. On 20 January 1921, only eight days before Mustafa Subhi and his associates were drowned in the dark waters of the Black Sea, the Turkish delegation was authorised to depart for Moscow. This time it was a larger delegation which contained four groups of advisers – plenipotentiaries headed by Yusuf Kemal, minister of the economy; a military advisory group led by Staff Major Saffet (Arikan); a political advisory group consisting of representatives from the Caucasus; and a study group headed by Dr Tevfik Rüstü (Aras).[96]

Less than a month after the 'Black Sea incident' the official talks between the Turkish delegation and the representatives of the Soviet government commenced on 26 February 1921. The negotiations were conducted on two separate subjects, one dealing with the political matters and the other involving questions of military assistance.

These were smoothly brought to a successful conclusion within a few weeks. The Turco–Soviet Treaty of Friendship and Brotherhood, signed in March 1921 consisted of 16 articles,[97] the principal provisions of which have been summarised as follows:

- The treaty gave Turkey both Kars and Ardahan. On the other hand Turkey accepted to hand over the suzerainty of Batum to Georgia. Mutual provisions were made for the exchange of populations in the ceded parts.
- The Soviet government accepted the abolition of the capitulations and neither party was to recognise treaties imposed by force on the other party.
- Moscow recognised the Ankara government with all the territories claimed in the 'National Pact'.
- With a view to guaranteeing the freedom of the Straits and their free passage for commercial purposes to all countries both parties decided to entrust the final elaboration of the regime of the Black Sea and the Straits to a special conference of delegates of the littoral states.
- Both parties agreed to accord the most favoured treatment to the citizens of each party residing on the territory of the other.
- The Soviet government accepted to undertake all necessary steps to secure endorsement of this treaty by the Transcaucasian Republics, in separate treaties which they were to conclude with Turkey, of such stipulations of the present treaty as relate directly to them.
- Each of the contracting parties agreed to take steps to prevent subversive propaganda by the Soviet citizens in Turkey and vice versa. It was further agreed not to allow on their respective territories the formation of groups that would lay claim to the role of government in the country of the other party.

As requested by the Soviet side, the issue of financial and military aid to Turkey was not included in the text of the treaty. This was a measure not to undermine their treaty with the British.[98] It was agreed during the negotiations that the amount and the nature of the Soviet financial and military aid should be finalised through the exchange of secret letters and that these letters should be considered an indispensable part of the treaty of friendship.[99]

The nature and amount of Soviet material aid is obscure.[100] In his speech Mustafa Kemal makes no reference to the Soviet aid. Halil Pasha, the first semi-official representative of the Ankara government in Moscow, was quoted to affirm that he had handed over to Cavit Bey, the division commander at Karaköse (a small town near Erzurum) some 100,000 Turkish liras–worth of gold bullion in May 1920 which was given to him by the representatives of the Soviet government.[101]

Cebesoy writes in his memoirs that Yusuf Kemal, on returning from his first Moscow mission headed by Bekir Sami, brought back to Ankara 1 million gold rubles in September 1920. It is further claimed that following the treaty of March 1921, Yusuf Kemal brought to Turkey 400,000 gold rubles and while he was in Moscow sent 100,000 gold rubles to Turkish agents in Germany with Staff Major Saffet for the purchase of war materials.[102] According to Selek, on the other hand, Major Saffet was given 1 million gold rubles, ten times more than the amount quoted by Cebesoy. There is no definitive information about the fate of this money which was sent to Germany. Selek claims that Major Saffet tried to increase the available money to purchase more weaponry, and used this money on the stock exchange. Instead of increasing it he lost it all.[103]

Sevket Süreyya Aydemir maintains that the Ankara government had requested from the Soviets 10 million gold rubles each year for the continuation of what they called the War of Independence[104] and that during 1920 approximately 5 million gold rubles were sent to Turkey. According to Aydemir, from 1921 to 1922, 10 million gold rubles were also received.[105] According to Karal, Soviet sources confirm the delivery of the following amounts of financial aid to Turkey: In September 1920 1 million gold rubles and 200.6 kilograms of gold were brought to Erzurum by Consul I. Upmal-Angarskii. Following the treaty in April 1921, Yusuf Kemal received 4 million gold rubles. During May and June 1921, Major Saffet was given 1.4 million gold rubles. In November 1921, General Mikhail Frunze (commander-in-chief of Soviet forces in the Ukraine) brought with him to Trabzon 1.1 million gold rubles. In May 1922, Ambassador Aralov brought with him to Ankara 3.5 million gold rubles. Karal further states that the minister of finance, Hasan Fehmi, gave to Ambassador Aralov a receipt for the full 10 million gold rubles upon the latter's delivery of the last instalment.[106]

A British intelligence report dated 21 November 1921 cites the following amounts of assistance received by the Kemalists from Moscow during the period of March 1920 to October 1921: 20 poods of gold (about 50,000 sterling) as the first instalment of a projected loan of 50 million gold rubles in July 1920. Two million gold rubles in November 1920. Forty million gold rubles, plus 1 million gold rubles sent to Turkish agents in Germany in April

1921. Finally, 500 okes of gold (about £90,000 sterling) in September 1921.[107]

There is not a complete list in the Soviet accounts, and the documents in the archives do not provide a totally consistent picture. According to some documents an agreement reached with Turkey during the period the Moscow Treaty was signed and a sum of 10,000,000 gold rubles given to Turkey.[108] Of this sum, 5,400,000 rubles were transferred to Turkey in instalments in April, May and June of the same year. Another 1,100,000 gold rubles were transferred to the Turkish government at the close of 1921.[109] On 29 December 1921, and 29 April 1922, the Soviet government turned over to the Turkish Consul at Novorossisk large amount of mines, artillery, rifles and other weapons. In 1922 Turkey was given equipment for a cartridge manufacturing plant. On 3 May 1922 the Soviet government transferred to Turkey 3,500,000 gold rubles which was the last instalment of the 10 million ruble loan.[110]

The information gathered from the Soviet archives and various Turkish accounts support the view that the money sent from Moscow to the Turkish nationalists never exceeded 10 million gold rubles.[111] The main disagreement, however, seems to be related to the nature of the agreement: while many Turkish accounts claim that the Bolsheviks promised to give 10 million rubles a year, the Soviet accounts insist that this was the total amount they agreed to pay.

The Turco–Soviet Treaty of Friendship has been the subject of much controversy and comment. For the Ankara government it was considered to be a major diplomatic achievement which secured a favourable position in its eastern frontiers. It strengthened its diplomatic and military position. Soviet historians claim that this treaty gave the Kemalists greater international prestige than ever before.[112] For Soviet Russia the treaty established in general its borders with Turkey in the Caucasus. It was also anticipated that this treaty would increase the prestige of the Soviet regime in the East and disperse suspicions among the Muslim population of the Caucasus. A few days after the signature of the treaty, in a speech delivered in Moscow, Lenin commented that 'a few days ago we signed a peace agreement with the Turks, which alone will rid us of interminable wars in the Caucasus'.[113]

Despite the general affirmative terms of the treaty as a whole, it can not be said that it indicated total trust on both sides. Article XV states that Russia would facilitate an understanding with the Transcaucasian states, yet in fact the Soviet government was not at this time keen to extend the agreement to the Transcaucasian Soviet states. Chicherin, in his telegram to Mamia Dmitrievich Orakhelashvili, chairman of the Georgian Revkom in Tiflis, urged those republics not to enter in any dealings with Turkey yet and to wait for directives from Moscow.[114] It was not until October 1921 that Turkey was able to effectively settle its eastern borders. By this time the Turks had fought a decisive battle against the Greek army near the Sakarya river in central Anatolia, which was considered as firm evidence of their authority in the eyes of Moscow. The extremely cautious nature of the rapprochement becomes most clear in article VIII where both parties warn off the propagandists of the other.[115]

Still, on a more general level, the treaty with Turkey, together with similar treaties with Persia on 26 February and with Afghanistan on 28 February,[116] constituted a further stage in the process of accommodation by which relations between the Bolsheviks and the outside world were consolidated on a governmental basis. Moscow regularised relations with its immediate neighbours, which it considered the foundation stone for a Soviet security system built upon close relations with key states in sensitive border zones.

5

Turkey Between East and West
March 1921–September 1922

We should realise that this is a propitious moment to bring about a settlement in Turkey, and it is vital to us that we up a buffer between our Empire and Russia. Unfortunately our withdrawal from Caucasia led to the occupation of these small republics by the Bolsheviks.[John de Robeck, British High Commissioner in Constantinople, 15 August 1921][1]

Although you are going officially to Ankara on behalf of the Ukraine, politically your visit will be considered as an expression of friendship of all the Soviet republics of Russia ... Therefore your visit will play a role to counterbalance of all those influences which pushes Turkey to the Allies' sphere of influence. The Turkish ruling circles will definitely listen to what you say with great interest, and will judge it as an expression of Russian policy ... Try to study in detail the state of the Turkish army, its control system and supplies, and the prospects of further operations. We must know to what extent the Turkish army remains and is bound to remain as an efficient, serious military factor ... We must know ...whether we should expect some surprises. You should estimate, from a military point of view, the general situation in Turkey, and the importance of such factors as ... the Kurdish rebellion and the activity among the Armenians and the Greeks. We must know not only whether Turkey remains as an efficient military factor but also whether there are any grounds to consider that it[the Turkish army] intends to turn against us as a result of an agreement with the Allies ... [From Molotov, Secretary of the C.C. of the RCP(B), to Frunze, 3 October 1921][2]

Mariage de convenance:[3] From the Moscow Treaty to the Turkish Victory in Anatolia

Although it was not exactly in the spirit of 'friendship and brotherhood' the Moscow Treaty of 1921 is an important point in the Turco–Russian relations. It was the first treaty in the history of relations between two countries which was not imposed upon one party by the military power of the other. It was the military power of a third party, of the British which brought Ankara and Moscow together. This common threat perceived as coming from the West, particularly from Britain, made it possible for both sides to sign an agreement despite their differences. Both the Soviet and Turkish governments expected this treaty to strengthen their international positions. Furthermore, such a diplomatic step on a predominantly governmental basis was expected to diminish the suspicions and animosities between both sides.

On 28 March 1921 according to the article XIII of the Moscow Treaty, the representatives of both parties signed a supplementary treaty in Moscow regarding the return of the prisoners of war back home.[4] The next day the Soviet government decided to send S. P. Natsarenus as their ambassador to Ankara.[5]

The Moscow Treaty was signed at a time of difficulty when the Soviet–Turkish relations had been strained following the competition on Armenia and general suppression of pro-Soviet and other radical groups in Anatolia. It would have been rather naive to assume that the treaty would easily bridge the huge gap of trust in relations.[6] The following months showed that the treaty of Moscow provided neither an end to the problems between two sides nor a promising fresh start.

Before the Turkish delegation returned to Anatolia the first trouble emerged which darkened the early optimistic mood of the post-Moscow Treaty days. According to the Moscow Treaty the Turks had promised to withdraw from Alexandropol. On 3 April A. I. Hekker, commander of the 11th Red Army, demanded from Karabekir the withdrawal of the Turkish forces from Alexandropol. Karabekir, however, wanted first the Armenian government to endorse the terms of the Moscow Treaty as a basis in its relations with Turkey.[7] In the meantime Chicherin handed a note of protest to Ali Fuat and repeated Hekker's demand.[8] On 13 April, Hekker sent a telegram to Karabekir stating that unless the Turkish troops withdrew the Red Army would force them to do so.[9]

Neither side, perhaps Ankara more than Moscow, desired this problem to escalate into a new crisis. It would not have been wise for both parties to fight over such an issue at a time when they were trying to present a united front in opposition to the Western Allies. 'Despite the difficulties, both sides had a good reason' to keep the relations at a reasonable level.[10] On 15 April the Turkish government in Ankara decided to evacuate Alexandropol in a week.[11] The last

Turkish soldier left the city on 23 April. A gesture came from the other side on the next day when the government of Soviet Azerbaijan sent a message to Ankara congratulating the Turkish government on the occasion of the first anniversary of the opening of the Grand National Assembly on 23 April 1920.[12]

On 25 April Yusuf Kemal, the head of the Turkish delegation, met Ordzhonikidze and Svanidze, Georgian foreign commissar. A decision was taken at this meeting to organise a conference at Kars regarding the agreement with the separate Transcaucasian republics.[13] By May 1921 the progress of mutual relations seemed promising. The Treaty of Friendship had been endorsed; the eastern borders were stabilised; and the Kars conference which would guarantee the recognition of the Moscow treaty by the Transcaucasian Soviet republics had been placed on the agenda.

It was not long before the emergence of the next dispute. Before the end of May Chicherin sent a protest letter to the Turkish representative in Moscow, concerning the maltreatment of the Molokans by the Turkish government. Molokans had been settled forcefully in the Kars province by the Tsarist authorities in 1896 since they had refused to do military service. Kars had been chosen as it was at the outskirts of the Russian sphere. It was the turn of the Ankara government, in its attempt to mobilise all available human resources against the Greek army, to impose military service upon the Molokans in 1921.[14]

Chicherin in his letter of 18 May protested against the recent situation and demanded a change in the policy.[15] Ali Fuat in his telegram to Ankara on 26 May criticised Chicherin's letter, saying that the Molokans were Turkey's domestic problem and the Bolsheviks were encouraging the Molokans to rebel against the Turkish authorities. Ali Fuat suggested whether the Bolsheviks might have been thinking of using this incident as an opportunity to make some territorial claims from the Turks.[16] On 10 June Mustafa Kemal took up Ali Fuat's concerns and urged Karabekir to take all necessary measures along the eastern border against a possible Soviet attack.[17] Karabekir, however, did not share this suspicion and considered all this 'fuss' as 'an Allied propaganda' that aimed to prevent the transfer of Turkish troops from the Eastern front to the West.[18]

A close examination of the Soviet documents shows that Karabekir was not wrong in criticising Ali Fuat and Kemal's suspicion. Indeed a military operation in the region against the Turks would be the last thing the Bolsheviks needed at that stage. There was another reason behind the anxiety and rising nervousness in Ankara. It was related to the latest movements of Enver Pasha, who had been in Russia since mid-1920 after his short exile in Berlin.

On 2 July 1921 the Greek army, which controlled Smyrna and surrounding area, initiated an offensive. By August they occupied Eskisehir, 140 miles west of Ankara, and started to advance towards the nationalist capital. In this precarious atmosphere and panic in Ankara, Enver was reported to be in Batum waiting to enter Anatolia to challenge Kemal's leadership in the national struggle. As the Greek advance became more threatening towards the middle of June 1921 there were indications in Ankara that Enver supporters were preparing a *coup d'état*. Natsarenus, who arrived in Ankara in June 1921,[19] wrote to Chicherin that there were preparations in Ankara by the supporters of Enver for an anti-Kemalist take-over.[20] According to a secret British Intelligence Report, however, Natsarenus himself was involved in the plot designed by Enver's supporters.[21] Soviet involvement in this conspiracy was confirmed by Ali Fuat who writes in his memoirs that when he met Chicherin in Moscow in mid-August Chicherin suggested that Enver should enter Anatolia at the head of an Islamic army to save the nationalists from a disastrous defeat.[22]

When the Turkish army halted at their positions and checked the advance of the Greek forces at the Sakarya river, less than fifty miles from Ankara, the Kemalist leadership hastily took some measures to consolidate its position. Due to the increasing morale at the nationalist camp the suspicion about Enver's intentions and the rumours about the role of the Soviets in whole affair gradually disappeared. Within this relatively relaxed atmosphere following the cease of the Greek threat over the nationalist capital, Natsarenus proposed on 20 August that the conference with the Transcaucasian Soviet republics should meet in Kars soon.[23] This was followed by a similar demand from the Armenian Soviet republic.[24]

The project of a conference to regulate the position as between the Caucasian states and the Ankara government had been under discussion for a considerable time before it actually assembled at Kars in late September.[25] Once the establishment of the Soviet power in Georgia had completed the process of the Sovietisation of the Caucasus, it was natural that steps should be taken to regulate various questions – such as the status of Batum and Kars *vilayets*. The agreement with the independent Soviet republics of the Caucasus was regarded as a suitable complement to the Turco–Soviet Agreement of March 1921. The conference at Kars opened on 26 September and lasted for about two weeks. The agreement was concluded between the Ankara government and the three Caucasian republics at Kars on 13 October 1921.[26]

While the proceedings of the Kars Conference were going on another high-level Soviet delegation was about to start its journey to Ankara. This time the issue was to sign a complementary agreement to the Turco–Soviet Agreement with the Ukrainian Soviet Republic. The delegation headed by Frunze left Moscow on 24 November and arrived in Ankara on 13 December.[27] Frunze settled into a round of negotiations with Yusuf Kemal with whom he signed a treaty on 2 January 1922.[28] This treaty – like those with the Soviet republics of the Caucasus – followed closely the text of the Moscow Treaty.[29]

Frunze left Turkey on 15 January from Trabzon. When he arrived in Moscow in late January, he presented his report on his visit to Turkey to the Central Committee.[30] His report described the financial difficulties of the Kemalist government in Ankara and urged the Central Committee to help immediately.[31] Kheifets claims that upon Frunze's request, the C.C. decided to send the remainder of its initial promise of 10 million gold rubles, which amounted to 3.5 million.[32]

As relations between Soviet Russia and Turkey were normalised in early 1922, the invitation of the Soviet government to an international conference at Genoa led the Turks to feel left out and suspicious. The Allied Supreme Council at its sixteenth meeting in the French Mediterranean resort city of Cannes in January gave its approval to a motion proposed by Lloyd George. This called for an international economic and financial conference in February or early March to which all the European powers, including Soviet Russia, should be invited to send representatives.[33]

The Soviet government saw Genoa as an opportunity for the normalisation of its relations with various countries. As for the Turkish affairs, the Bolsheviks tried to use this opportunity to secure their share of control of the Black Sea and the Straits along the line parallel to article V of the Turco–Soviet treaty of March 1921. Article V states:

> In order to assure the opening of the Straits and the free passage of merchant ships of all nations, the contracting parties agree to entrust the final elaboration of an international regime of the Black Sea and the Straits to a future conference composed of delegates of the littoral states, on condition that the decisions of the above-mentioned conference shall not inflict damage on the full sovereignty of Turkey or on the security of Turkey or of Constantinople, her capital ... [34]

This article in fact proposed a return to the pre-war status of the Straits and showed that although it supported commercial freedom in the Straits, the Soviet government was opposed to warships entering or leaving the Straits. This was against the policy of the Allies who wanted the Straits to remain open to vessels of commerce as well as war, as had been one of the central provisions of the Treaty of Sèvres. The Straits were still under Allied control and the Soviet–Kemalist treaty had no immediate relevance. The Soviet government decided to use this opportunity to bring the issue to the agenda of an international conference at Genoa.[35]

Turkey was not invited to the Genoa Conference. The Allied governments wanted it to be essentially an economic conference which would include all the nations of Europe and the United States. The Soviet government, on the other hand, tried to secure an invitation for Turkey. Chicherin asked Krasin, the Soviet trade delegate in London, to do everything possible to secure an invitation for the nationalist Ankara government.[36] Krasin appealed to the British Foreign Office 'to use its influence towards obtaining the participation of the Turkish government in the Genoa Conference'. He stated that 'Russia is specially anxious that the question in dispute among the countries of the Black Sea, and especially the question of the Straits, should be settled at the conference ...'[37]

The British, on the other hand, had suspicions regarding the Soviet demand about Turkey's inclusion to the conference. The S.I.S. alarmed the Foreign Office by saying that a key Russian demand would be that the Straits of Constantinople given to them. Leeper (Northern Department of the Foreign Office) showed his opinion on this by not forwarding it to Genoa.[38] In the meantime Krasin was informed by the British Foreign Office that the conference would 'in no way be concerned with the Turkish question, and in any case the question of the issue of an invitation to the Turkish government to participate in such a conference does not appear to be one in which the Russian trade delegation in London or the Soviet government in Moscow are concerned'.[39]

Despite this official line, 'the Genoa Conference and the struggle to control Anatolia and the Straits were interconnected'.[40] The British government was not willing to include the Turkish question in the official agenda of the conference but there was a strong intention within the British circles to use the Genoa Conference as an opportunity to secure the co-operation of the Soviet government 'for the purpose of establishing peace in the Near East'.[41] It was stated in a Foreign Office memorandum that 'so long as they [the

Kemalists] have the sympathy of Russia they can rely on her for military help' and resist pressure from the Allies.[42] The British held the firm belief that 'a settlement in the Near East is contingent on a settlement with Russia'. It was hoped that as a result of Genoa it would be possible 'to make Russia neutral in her attitude towards Ankara'.[43] This point was clearly emphasised in a draft agreement prepared by a small drafting committee in the Foreign Office at the request of the Interdepartmental Committee set up to consider preparations for the Genoa Conference. The undated draft, presented to Sir E. Crowe on 8 March, demanded the contracting parties 'to refrain from propaganda subversive of the order and established political system in another's territories'. The draft agreement further demanded 'the Russian Soviet government ... to adopt an attitude of strict neutrality towards Greece and the Grand National Assembly at Angora [Ankara] and to cease all state assistance in money and munitions to either party.'[44] Furthermore, in Genoa the British openly asked the Soviet representatives 'to hinder the Angora [Ankara] government from committing war-like operations'.[45]

In Ankara, however, the invitation of Moscow to the Genoa Conference and the exclusion of the Turkish government caused great concern and panic among the Kemalists. They interpreted these developments nothing but an Allied attempt to separate the Bolsheviks from the nationalists.[46] When Chicherin signed the treaty with Rathenau[47] in the small seaside village of Rapallo on 16 April 1922 it convinced many that this 34-nation summit conference was going to be a failure, at least in terms of its immediate results.[48] Mustafa Kemal remarked that 'this is a great victory for Russia against the British. It has really given me much pleasure.'[49]

After the Genoa Conference collapsed on 19 May without achieving a general agreement, Karakhan sent a letter to Mustafa Kemal informing him that the Soviet delegation had been asked at Genoa to adopt a neutral position in the Near Eastern conflict and stop helping the Ankara government. Karakhan confirmed that the Soviet delegation had criticised the conference for not inviting Turkey and repeated the Soviet government's firmness in its friendship with Ankara.[50]

Following the Genoa Conference relations between Ankara and Moscow were kept at a stable level. There was no reason to suppose that Moscow and Ankara regarded each other with anything but profound mistrust but there was equally no reason to think that a complete rupture of relationship could occur before they would settle their respective scores with the Western powers. The '*mariage de convenance*' between the Kemalists and Bolsheviks was maintained and the transfer of armaments and ammunition to Anatolia continued regularly.[51]

One persistent thorn threatening to undermine the mutual understanding was resolved during the summer of 1922. Enver Pasha, whose actions in Russia had long been the source of much anxiety and concern, was killed in August 1922 whilst fighting against the Red Army troops in Central Asia. The victory over the Greeks at Sakarya River (August–September 1921) had upset the plans of Enver and diminished the hopes put upon him by Moscow. Enver Pasha, losing all his hopes to take over the national movement from Mustafa Kemal, had then left the Caucasus for Moscow. There he had been persuaded by the Bolshevik leaders to proceed to Turkestan. This vast country, bordering on China, India, and Afghanistan, the ethnic cradle of the Turkish people, was torn by a number of revolts that had brought anarchy to the emirates and kingdoms formerly ruled by the Russian Empire. The Bolsheviks were having great difficulty establishing order there. Moscow decided to send Enver to Bukhara to mediate with the rebel chiefs.[52]

Truth and legend about Enver Pasha are so inextricably mixed that it is difficult to be sure where one ends and the other begins. In his memoirs Roy says that the Soviet Political Police (GPU)[53] had sent him [Enver] to Central Asia 'to get him out of way':

> Karakhan was fed up with the continued presence of unwelcome guests in his courtyard. ... but it would be an international scandal if Enver Pasha simply disappeared from Moscow. ... let the embarrassing guest go to remote Central Asia, ...[54]

This claim is supported by Essad-Bey who writes in his book (*Secrets of the OGPU*[55]) that 'the Armenian Agabekov' – a young Intelligence officer who was trained as an expert in Oriental affairs – 'was sent to Bokhara [Bukhara]' for the special task of preparing ground for the liquidation of Enver.[56]

According to Karabekir in Bukhara one of Enver's supporters, Haci Sami, persuaded him to join the rebels, who were known as the *Basmaci*,[57] and lead their struggle against the Bolsheviks.[58] Enver, who only a few months before had talked of the necessity of a united Islam fighting beside the Soviets against the western imperialists, announced in Bukhara that the real oppressor of Islam was 'Red Imperialism'. He declared that his aim was to unify the different *Basmaci* groups operating in eastern Bukhara and form an army out of them to force the Soviets out of Turkestan.[59]

Enver Pasha commenced operations in early spring by organising several detachments of malcontents on the River Surkhan. At the same time auxiliary forces subservient to him were formed in the region of Bukhara. *Pravda* wrote that an attempt was also made by Enver to organise the *Basmaci* in Fergana and Samarkand.[60] On 19 May 1922 he sent an ultimatum to the Soviet government, via Nariman Narimanov – the president of Soviet Azerbaijan – demanding the recognition of the independence of Bukhara, Turkestan and Khiva by Moscow and the evacuation of all Turkestan by the Soviet troops. He gave Moscow two weeks to withdraw the army and remove the Russian administrative bodies that were functioning there.

In the meantime, the Bukhara Red Army dealt with the insurgents near Bukhara. In June, the Red Army began a movement to encircle Enver and his forces. On the night of 24 June Enver attacked the Soviet forces in the region of Derbent (in Uzbekistan) and was defeated. In July the Soviet forces reoccupied Dushanbe and Enver's partisans abandoned him *en masse*.[61] Enver, having lost practically everything, with a handful of men, retreated back to the mountains on the Afghan border. On the morning of 4 August 1922 a Red Army patrol appeared in the mountain gorge. Enver was killed after a short battle with the much stronger Red Army patrol.[62] He was heading a platoon-sized force, and was reported to have died sword in hand assaulting a machine-gun position.[63]

All his life Enver Pasha had thought in terms of great ideals, historic missions, and symbolic sacrifices. All his strivings were aimed towards changing the course of history in accordance with his ideals. He gave himself unrealistically grandiose titles. At the Baku Congress he was 'the representative of the revolutionary organisations of Morocco, Algeria, Tunisia, Tripoli, Egypt, Arabia and India'. In Bukhara he signed himself as 'Commander-in-chief of all the armies of Islam, son-in-law of the Caliph and representative of the Prophet'. To his wife Naciye Sultan, he wrote on 7 November 1921 of having signed proclamations to Bukhara and Khiva as 'the Commander of the great Turanian Revolutionary Armies, Supreme Head of the Central Organisation'.[64] Enver was the prisoner of his idealism. He was 'a great idealist, unable to reckon with life and with the realities.'[65] When he realised the impossibility of putting his ideals into practice, 'he threw himself into the arms of death, willingly, deliberately.'[66]

With Enver's tragic end in August 1922, an important source of suspicion in Turco–Soviet relations disappeared on its own account.

However, this was not accompanied by a distinct improvement in relations. During the summer months of 1922, when Enver Pasha was approaching his final battle, Ankara witnessed a series of political developments which set a pattern indicating a slow but significant shift in the orientation of the Turkish foreign policy.

From the beginning of Turco–Soviet relations, the central factor in Ankara's approach to the Soviet government had been the military threat stemming from the operations of the Greek army supported by the British. As Soviet Russia was the only external power willing to provide Ankara military help and diplomatic assistance, relations between Moscow and Ankara had been given top priority. But when the Turkish army succeeded in stopping the Greek troops at Sakarya River, Turkish nationalist self-confidence rose to an unprecedented level. The Kemalists realised their advantageous bargaining position and felt less dependent upon Soviet help than before.

Within this new and most promising atmosphere following the Sakarya battle, the ardent critics of the relations with Moscow gradually started to gain ground in Ankara. This change in the nationalist camp was reinforced on 12 July 1922 with the appointment of Rauf Bey, who had always been the most ardent critic of Turco–Soviet rapprochement in the Grand National Assembly, as the prime minister. The next day a close associate of Rauf, Adnan Bey [Adivar], became the deputy head of the parliament.[67]

These two appointments were accompanied by a series of legislative and other administrative steps – all of which aimed at a more pronounced anti-communist policy.[68] The new cabinet banned all communist and socialist propaganda. Magdeleine Marx, the granddaughter of Karl Marx who visited Ankara in the months of July and August 1922, describes a pungent anti-communist and anti-Soviet atmosphere in the Kemalist capital.[69]

On 15 August the Soviet ambassador's residence at Ankara was burnt down. Aralov, the ambassador, suspected an arson attack and asked Moscow to review its Turkish policy. He also advised a sharp curtailment of supplies until Ankara provided some explanation to the recent anti-Soviet incidents.[70] Aralov in his memoirs puts the blame on Rauf Bey who is said to have advocated anti-Soviet policies for a long time. Now, with the authority of the premiership, he was taking an actively hostile stand against the Bolshevik regime during the summer of 1922.[71]

Aralov links this change in the attitude of the Ankara government to an international conspiracy. According to Aralov it accounted for the increasing influence of French diplomacy among the Kemalists.

Indeed, on 4 June, about two months before the suspected arson attack at Aralov's residence, a representative of the French government, Colonel Mougin, came to Ankara to discuss the possibility of further collaboration between the two sides. The Turkish press described the main aim of the French mission as being to investigate the Bolshevik influence in Anatolia and counteract the high profile presence of Aralov in Ankara.[72] Aralov shares this view and claims that Mougin had a secret mission. Mougin, Aralov believes, approached Mustafa Kemal with a proposal to sign a military alliance with France, to make a compromise peace with Britain, and to sever his relations with the Soviet government.[73]

Soon after the Sakarya Battle the Turkish foreign minister Yusuf Kemal had signed an 'accord' with Franklin-Bouillon, a Turcophile Senator and former president of the Foreign Affairs Commission in the French Senate. This agreement (known as the Ankara Accord and signed on 20 October 1921) had promoted the nationalist policy of splitting the Allies by dealing with each one separately. Mustafa Kemal freed his armies in Cilicia after the Franco–Turkish Agreement.[74] This was followed by further indications of Turco–French co-operation in the region – such as importing war material from France.[75]

The government re-shuffle and the introduction of a ban on communist propaganda indicate that the suspicions of Aralov were not fully unjustified. However, it is difficult to confirm the real significance of the French attempts in bringing about such a U-turn in the nationalist policy. What is certain is that in the second half of 1922 the Kemalist movement was not as desperate as it had been previously to ensure Soviet support.

British–Turkish Relations from Early 1921 to September 1922

During the same period the British had also initiated a number of attempts to find a solution to the crisis in Anatolia and to diminish the Soviet influence on the nationalist movement. While the Kemalists and the Bolsheviks were meeting in Moscow for the last round of the negotiations towards the Moscow Treaty in the winter of 1920–21, there was increasing support on the British side for the review of their policy towards the Turkish national movement. Lloyd George in a statement in the House of Commons on 22 December 1920 stated that 'I want to act justly towards the Turks and I am ready to discuss the proposals submitted to me by the Turks'.[76] This statement can be traced back to the decision of the Commission of

British Foreign Affairs on the Turkish problem which met on 20 December and came to the following conclusion:

> The Commission draws the attention of Mr Lloyd George to the Turkish situation and the Commission hopes that the necessary steps will be taken to re-establish a regular government in Turkey and, if opportunity occurs, to modify the Treaty of Sèvres thus separating the Nationalists from the Bolsheviks.[77]

In the early days of 1921 further signs appeared on the British side which indicated a softening of the attitude towards the Turks. General Harington was writing in mid-January to the War Office that the Turks were 'outstanding elements on the side of law and order', and 'by financial assistance and sound administration Turkey can be placed on her legs'.[78] In late January Horace Rumbold was writing to Curzon that it was no use to regard Mustafa Kemal any longer as a 'brigand chief'.[79] According to Rumbold the Ankara government had already proved that they had an effective control and full functions of government over the whole of Asia Minor.[80]

On the same day Rumbold sent his telegram to Curzon, the British Cabinet took up the Turkish problem and came to the decision that some modifications in the Turkish treaty were necessary as long as its main lines were kept intact.[81] On 24 January Lloyd George and Curzon went to Paris to discuss the issue with the French and the Italians. There it was decided that some modifications of the treaty should be offered to the Turks. The main concern in the Allied camp was clearly related to the rising profile of the Bolshevik influence in Turkey. It was stated that the 'unnatural and dangerous' alliance between the Turkish nationalists and the Bolsheviks could only be prevented by a more conciliatory approach towards the Kemalists.[82]

In February an invitation was sent out to the Greek government, the Sultan's government in Constantinople, and to the Ankara government to attend a conference in London to discuss the terms of the Treaty of Sèvres. This invitation was significant as the Allies, by inviting the Ankara government for the first time to a conference with the objectives of ending the Turco–Greek war and of examining the ways and means of revising the Turkish treaty, *de facto* recognised the authority of the Kemalist government.[83]

Although the Turkish nationalists did not expect too much from the conference, they decided to send their delegation. Mustafa Kemal, in a secret sitting of the Grand National Assembly on 12 February 1921, described the proposed Allied conference in London as follows:

> It is common knowledge that Anglo–Russian rivalry is entering on a new phase in the East. For the purpose of hastening the conclusion of an alliance with the Russians, the English have invited us to the Conference of London and we have sent our delegates there … I am not placing great hopes in it. I am not sure of the good faith of England … Russia wishes to found a Soviet government in Anatolia and I have learned that of late the Russians have financed various committees in the interior of Anatolia … we are now engaged secretly in tracing them down … Our national government, however, has seriously considered all these points and adopted the necessary measures.[84]

Bekir Sami, who had just returned from negotiations in Moscow, led the Turkish nationalist delegation at the London Conference. At the conference he spoke for the Constantinople government as well as the Ankara government – an indication of both the harmony and the relative power of Constantinople and Ankara. Bekir Sami seemed to be willing to consider, with reservation, some of the modifications of the Sèvres Treaty suggested by the Allied powers. He submitted a declaration incorporating the minimum terms for a political solution. These were basically a reiteration of the National Pact. The declaration claimed full sovereignty rights over the non-Arab inhabited territories of the Ottoman Empire.[85]

The Ankara delegation tried to convince Lloyd George that a Turco–British understanding could be beneficial to both parties.[86] While another Turkish delegation was desperate to ensure the Bolsheviks' support in Moscow, Bekir Sami was anxious in London to describe the relations with the Soviet state as temporary and something they would be gladly rid of if they could come to terms with Britain.[87] Bekir Sami, himself a native of the Caucasus, proposed an alternative: Turco–British co-operation to create a federation of independent Caucasian states that would not only contain Bolshevik expansion but also, by depriving the Soviet government of considerable food and fuel resources, bring them to their knees and put them at the mercy of Britain.[88]

The negotiations in London lasted from 23 February to 12 March. On 16 March 1921 (again the same day as the British–Soviet and Turco–Soviet agreements were concluded) Bekir Sami signed an agreement with Robert Vansittart of the Foreign Office for the release of all British prisoners of war and other British detainees held in Turkey in exchange for Turkish prisoners of war and interned civilians in the hands of British authorities.[89] Bekir Sami at least secured the release of nationalist leaders arrested in Constantinople during the reinforced British occupation a year previously.

The London Conference failed to find a solution to the Turkish question acceptable to all parties. The gap between the Turkish nationalists' demands and what the Greeks were prepared to concede was unbridgeable. The Allies – each for their own reason – were unwilling to force a solution at that stage. They could go no further than to mediate and they could not prevent the resumption of hostilities. Without expressing agreement the delegations returned to Athens, Constantinople and Ankara respectively. The only party to gain from the conference was Ankara, which had received its first recognition from the Allies as the *de facto* representative of Turkey.

After the London Conference had broken down in resentment, the need for a show of strength and resolve via indisputable military victory seemed the only way for the Greeks to demonstrate their control of the situation. The Greek army in Asia Minor had been asked previously whether an operation, to begin before the opening of the conference, was possible. The Greek chief-of-staff replied that a new offensive to effect the permanent occupation of the Izmit–Eskisehir–Afyonkarahisar line could be contemplated – but not before April.[90] In the course of the discussions in London the Greek government was more interested in gaining time while the army on the ground was strengthening its defences. Greece regarded victory on the battlefield as the only viable option.

The Greek army moved on 23 March 1921 and first encountered only small Turkish units. The Turkish forces were driven out of Afyon Karahisar on 27 March. The Turkish resistance stepped up 35 km north-west of Eskisehir where the Turks maintained well-entrenched positions. This was 'the site of Söyüd [Sögüt], the first Anatolian village possessed by the ancestor of the Ottoman Dynasty, and now the Osmanlis were fighting for their national existence on the very spot where that existence had begun', writes Arnold Toynbee, reporting on the war front for the *Manchester Guardian*.[91]

Three successive Greek attacks were thrown back and the Greek army broke off the offensive. Fierce clashes followed with heavy casualties on both sides. By 10 April the Greek forces were back to the fortified positions from where they had started on 23 March, having narrowly escaped a disastrous defeat.[92] Following the resumption of open hostilities by Greece on 23 March, the British government had declared its neutrality in the Greco–Turkish conflict.[93] When the Greek advance was halted in early April the British military leaders were quick to project the implications of the failure of the Greek spring offensive. Wilson, the C.I.G.S., had described the General Staff's apprehension over the Greek operations as 'an

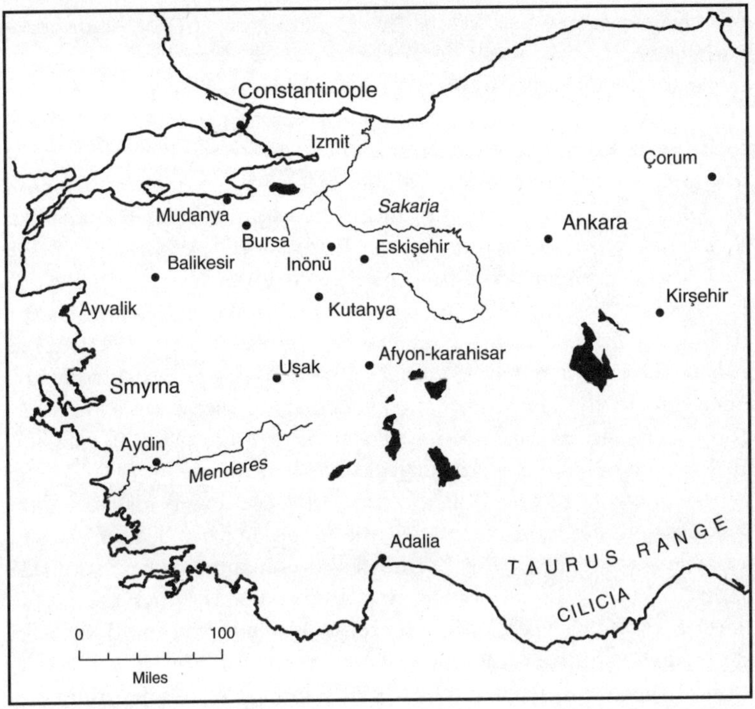

Map 6 Western Asia Minor

understatement'. If the Greeks collapsed, he maintained, it would become impossible for the British forces to keep their positions around Constantinople and the Straits.[94]

After the London Conference and following the Greeks' unsuccessful spring offensive, the Turkish nationalists felt stronger both in a military and political sense. This compelled them to a more determined stand against the Allies. In the aftermath of the newly signed Friendship Treaty with the Soviets general feelings towards the British were at an all time low. During the summer of 1921 the Ankara government adopted a more militant anti-British policy. A number of incidents contributed to the worsening of the relations.

In early May an Indian-born British citizen, Mustafa Saghir, was arrested by Kemalist authorities in Ankara.[95] Mustafa Saghir had arrived in Ankara as a delegate of the Peshawar Khilafat Conference to express Indian Muslims' sympathy for the cause of the Turkish national movement.[96] He was tried on a serious charge of espionage. During the proceedings of the trial Mustafa Saghir admitted the existence of a plot. He explained the details of pro-British propaganda being carried out by British agents in Constantinople and Anatolia. The accused further gave the names of prominent English officers who were chief parties in that scheme. He also mentioned a special committee set up to arrange the assassination of Mustafa Kemal. The Kemalist authorities reported that important documents and other evidence, incontestably establishing the guilt of the accused, had come into their possession and this had enabled them to expose the identity of the British spy.[97] At the end of the trial Mustafa Saghir was found guilty and immediately executed.[98]

This incident was followed by other minor skirmishes such as the raiding by the nationalist militants of the British ship 'Palitana' at Adalia, the closing down of Abbot's Emery Mines (a British company operating in the Italian zone), the denial of access to nationalist-controlled port of Scala Nova of the British steamship 'Elpiniki',[99] and the increasing anti-British tone of the Ankara press.[100] To the British the role of the Bolsheviks in all these anti-British activities was obvious. London interpreted the increasing hostility of Ankara as a clear evidence of the Bolsheviks' dominance in the nationalist capital. Several British sources reported a marked increase in the number of Bolshevik agents. According to one report the Bolsheviks succeeded in introducing their agents even into the Allied police, clerical, staffs and post-offices in Constantinople.[101] When the news of Mustafa Saghir's arrest arrived in London the British authorities tried to find out whether it was the work of the Soviet intelligence organisations which had helped Ankara to reveal the true identity of the British agent.[102]

This state of affairs caused discussion in London of ways to improve the profile of Britain in Turkey. Curzon considered options which would make a strong impression on the Kemalists. He suggested the raising of the embargo on the export of arms and ammunition to Greece. General Harington and Admiral Webb suggested a general naval blockade of all nationalist ports. This was to be enforced in such a way as not to provoke conflict with the Soviet government.[103] While the British were discussing the merits of a harsher policy on the Anatolian conflict, the Allied representatives who met at Paris in the middle of June decided to renew their mediation between the belligerents. The new terms were, briefly, an extension of the London proposals. In addition, this time they included a proposal about the Caucasus – mainly intended to pit the Turks against the Bolsheviks in the region. It might have been a diplomatic manoeuvre to balance against the recent success of the Bolsheviks in the Caucasus. With the seizure of Georgia on 18

March 1921 the Soviet government had definitely pulled the Caucasus within the Bolshevik sphere of influence. When the Allies met in June 1921 they showed an concerted eagerness to challenge the Bolsheviks' authority in the region, perhaps as a last chance. A number of concessions and some assistance to the Kemalists were expected to be enough to encourage them to expand in the Caucasus at the expense of the Bolsheviks.[104]

It was agreed that the new set of proposals would first be submitted to the Greek government. Negotiations with Constantinople and Ankara could then follow subject to Greek approval. If the Greeks turned down the proposal, then the initiative would be allowed to die. Athens, however, unaware of the Allied plans, was in a state of euphoria. The army leaders were confident of a comfortable victory in Anatolia. And also Rizo-Rangabe, the Greek chargé d'affaires in London, had been giving his government a totally wrong impression of the British position in the conflict, and the Greek government believed that they still had an open cheque.[105] It duly refused the Allied proposal. This refusal of Allied mediation meant a fresh wave of hostilities. In July 1921 the Greeks decided on a new offensive with the intention of pursuing the Kemalists to Ankara.[106]

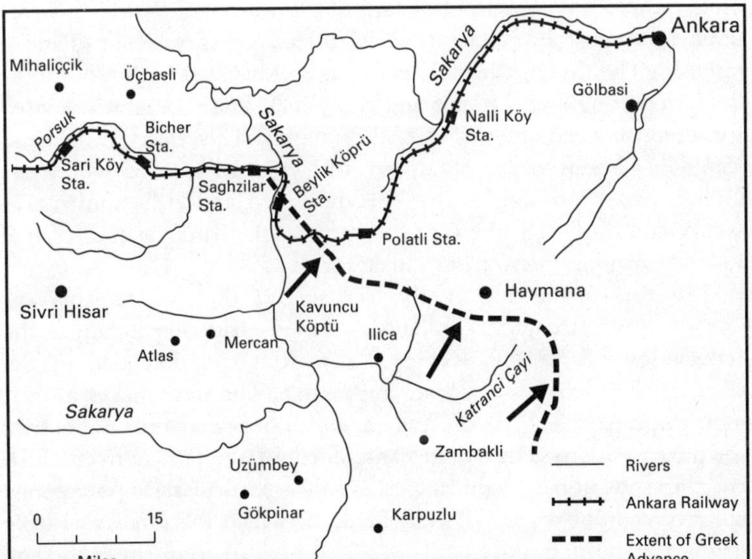

Map 7 The Sakarya Battle

On 11 July the Greek offensive began on the Smyrna front. The Greeks at first advanced practically unresisted as far as Afyon Karahisar, which was reported to have been occupied on 13 July. With the occupation of Kütahya and Eskisehir, the nationalists decided to evacuate the capital and to move to Kayseri (another central Anatolian town 160 miles south-east to Ankara). By the end of July the Greek forces reached Sivri Hisar – some 60 miles short of Ankara. Panic gripped Ankara and an outcry rose against Mustafa Kemal's leadership.

The easy advance of the Greek forces led the Greek High Commissioner to believe that the retreat of the nationalists would soon result in an anti-Kemalist revolution in Anatolia. Rattigan, on the other hand, feared this would take the form of a Bolshevik–Enver combination with the object of continuing the war to the bitter end. He advised the Greeks to show moderation and make it easier for the moderates in Ankara to take charge and conclude a peace.[107] The Greek government, however, believed that the complete break up of the Kemalist movement was imminent. They decided to continue the campaign until the enemy was completely defeated.

On 13 August 1921 the Greek army started its advance towards Ankara across the arid steppe land of Anatolia – the most difficult country it had yet encountered. The main Greek force advanced only 10 miles in as many days and did not meet the Turkish forces until 23 August by which time its effectiveness was considerably reduced. The Greek advance had in fact exhausted itself. Its supply lines were long and difficult and the Greek army was not prepared for desert-like conditions in high summer. This dusty, stony and waterless plateau was nothing like the western Anatolia where the Greeks had been able to live off the green and rich countryside. Weary and dispirited, the Greek army met the Turks at the Sakarya River – 40 miles west from Ankara.[108]

The Turkish position on the east side of the river Sakarya was one of great strength. The Turkish troops had dug in under the supervision of Mustafa Kemal. Their front was defended by the river itself which was tactically impassable. The river makes a great loop some 50 miles west of Ankara, which provided a natural barrier for their positions. Their right flank, north of the Porsuk river which cut their position at right angles, lay in a particularly mountainous country without roads. The left flank, by which the Greeks actually made their turning movement, was bent back to conform to the line of the river and, being in desert country without natural features advantageous for defence, was carefully strengthened by entrenchments. Behind the Turks was the Ankara railway which was vital for supplies and reserves.[109] In addition to their advantageous position, there was the hardy Anatolian peasant who made an ideal infantryman – may be not ideal on the attack position but who was determined to stand, fight and die if necessary. This was precisely what the nationalist leadership required.[110]

After eight days of serious fighting and heavy losses on both sides, neither side had obtained a decisive success. The two armies had fought themselves to a standstill. Both were short of ammunition and had suffered enormous casualties. On 9 September the Turkish army was just capable of one more counter-attack, the Greeks were not. Kemal, assuming personal command, made a daring enveloping attack with a relatively small force against the Greek left. On the evening of 11 September the Greek army started to retreat westwards across the Sakarya River. Soon it fell back along the railway line on intermediate entrenched positions covering the Smyrna–Aydin province.[111]

The battle of Sakarya represents a decisive turning-point in the Greco–Turkish war. In Toynbee's words:

> The Greeks were finally compelled to give up and with broken morale at last were driven back ... It was not a conclusive success [for the Turks], for the Greek Army was not destroyed ... but ... it was in fact, the turn of the tide in the Graeco–Turkish war.[112]

In military terms there was indeed a stalemate. Nevertheless, the Turks had achieved a political victory. The psychological effect of the Turkish move was immense. The battle of Sakarya broke the offensive capacity of the Greek army and forced it to take a defensive posture. With their 350-mile line of communications threatened, and the Anatolian winter coming on, the Greeks disengaged. In Churchill's words, the Greeks had involved themselves in a politico-strategic situation where 'anything short of a decisive victory was defeat'; and the Turks were in a position where 'anything short of overwhelming defeat was victory'.[113]

According to Comintern sources, the Greeks, after the failure of the Sakarya campaign, tried to find a diplomatic solution which would make them able to evacuate Anatolia without losing their face. The Comintern further commented that the military fiasco at Sakarya politically put the future of Greece in the hands of the Allies more than ever.[114]

The period following the Sakarya battle, particularly the winter months of 1921–22, was a period of extreme hardship for the Greek forces in Asia Minor. They remained in their positions in western Anatolia watching anxiously for the Turks next move. Under no circumstances were they ready to take the initiative. The Greek army was short of supplies, troops were frequently unpaid, there were a high number of desertions, and, most important of all, morale had sunk dramatically. The British authorities were anxiously watching this situation – being afraid of that the Greek army 'would soon cease to exist as a fighting machine'.[115]

Even before the decisive battle of Sakarya the British government had officially declared its neutrality in the Turco–Greek war.[116] But it was impossible to ignore the fact that British influence and prestige were still closely linked to the fate of the Greek army. The British were no longer the allies of the Greeks. They supplied no arms, no money or assistance and, most important of all, they had no control over the decisions of the Greek king, his politicians or his generals. However, the Greeks were still convinced that in an emergency the British, or at least their leader, would come to their help. In Churchill's words, the Greeks still believed that 'the Great Man is with us, and in his own way and in his own time and by his own wizardy he will bring us the vital aid we need.'[117] It was a dangerous situation for the Greeks, and was to prove equally dangerous for the 'great man' himself, the British prime minister, Lloyd George.

Demand for a revision of the British government's position towards Turkey grew in the aftermath of the battle of Sakarya. This now seemed even more urgent in the light of the recent Bolshevik operations in the Caucasus. At least they still had the opportunity to secure Anatolia from Bolshevik influence. Even before the final collapse of Georgia and Armenia, Hankey (the cabinet secretary) had written in his diary that a revision of British policy in favour of the Turks would please the Muslim world and reduce Britain's troubles in India and Egypt.[118]

A British report on the nationalist foreign policy shows the belief that the time was right for a sympathetic approach to the Ankara government. The report reads as follows:

> The exchange of prisoners of war, which has just been affected, has produced a favourable impression in Anatolia ... If they [the Kemalists] can only obtain an assurance from Great Britain that we [the British] will not lend the Greeks any moral or material support, they [the Kemalists] are perfectly willing to make terms with us ... Should England refuse to come to terms with Angora, the Nationalist government will not hesitate to use all means in their power to carry out anti-British propaganda throughout England's Muslim territories. If England will agree to rational

> terms, the Nationalists can practically assure her peace in Egypt, Mesopotamia and India.[119]

The arrival of a high-level Soviet delegation led by Frunze in Ankara in early December 1921 increased pressure upon the British government to adopt more conciliatory policies towards Ankara. It was seen necessary to counter-balance the Bolshevik influence in the Kemalist camp.[120] In mid-December 1921 Curzon proposed a new meeting, which met with approval from the British Cabinet.[121] Curzon urged the Allies to go to the 'utmost limit of reasonable concessions' in order to reach a settlement with the Turks.[122] The same day a telegram arrived from Constantinople which provided further support for Curzon's case. According to it the Ankara government, following the battle at the river Sakarya, had decided to carry out an offensive in the Afyon Karahisar area during November. But a British report informed the government that the offensive had been postponed due to the Kemalists' concern that it might offend Britain as 'it was thought that irritation might be caused in British circles if active operations were undertaken against the Greeks'.[123]

The meeting of the Allies, which was proposed by Curzon in late 1921, did not take place until 22 March 1922 in Paris. Allied powers agreed that, in the interests of the peace and security of the whole region, the highest priority should be given to weaning Turkey away from Moscow. The Allies were confident that now both Greeks and Turks were ready for a fresh round of peace initiatives. The military stalemate and the complete exhaustion particularly of the Greek army were considered to provide a favourable ground for this initiative.[124]

The Allies agreed to propose first an armistice to the governments of Turkey and Greece.[125] The announcement of the terms of the proposed armistice came at a time when the Greek government, and the Asia Minor Army in particular, was having great financial and military difficulties. Gounaris, the acting prime minister at Athens, had described the desperate situation of his government in a letter to Lloyd George on 27 February, 1922. In his letter Gounaris had confirmed that:

> The complete exhaustion of our financial resources has made the possibility of maintaining the Army in Asia Minor problematic. In addition to this, shortage of war-material must necessarily expose the army to grave danger as soon as the campaigning season sets in.[126]

This time the Greeks immediately accepted the armistice offer subject to a few minor reservations.[127] Having failed to carry out their 'great idea', they felt that they had to rely on Britain to bail them out of Anatolia.

As soon as the terms were notified to Ankara, Mustafa Kemal gave a copy of the armistice proposal to the Soviet envoy, Aralov. Aralov writes in his memoirs that Mustafa Kemal described the Western terms as 'condemnation to hard labour.'[128] However, the Turkish government did not consider it tactful to refuse the proposed armistice as they did not wish to be the party responsible for a rupture. Ankara accepted the armistice in principle with counter-proposals, demanding the evacuation of Anatolia as a first condition.[129] This reply, however, did not satisfy the Allies.[130] The mediation attempt came to a standstill as before.

The Turkish refusal of an armistice on the basis of their demand for the Greeks to evacuate Anatolia was criticised sharply by Lloyd George in his speech to the House of Commons on 4 August. He put all the blame for the Anatolian conflict on the Turks, while encouraging the Greeks 'to wage war with their full strength'. He stated that 'There is only one way in which the Greeks can have a decision ... and that is by marching through almost impenetrable defiles for hundreds of miles into the country ... '[131]

The Ankara government decided to force a last attempt to win over the British. In July Kemal decided to send his minister of the interior, Ali Fethi, to Europe on an unofficial visit to the Allied capitals to hear the views of the entente statesmen. Fethi was given complete authority with the full confidence of Kemal to initiate talks and to negotiate peace.[132]

In London Fethi talked to Lord Long, an acquaintance of Curzon. He repeated the Turkish position as being the complete demilitarisation of the Dardanelles with absolute freedom of ingress, the establishment of a scheme for the full protection of the minorities and the removal of the Greek frontier from the line of Edirne to the mouth of the river Maritza. On 14 August, with Curzon's authorisation, Lindsey and Tyrrell of the Foreign Office met Fethi and discussed the Turkish peace offer with him. Five days later, however, Lindsey informed Fethi that he would not be given permission to stay in London any longer. Fethi then reported to Ankara that he was convinced that the British prime minister and his foreign minister were set on the destruction of the independence of Turkish homeland. He felt that due to the uncompromising attitude of the British there remained no way for the Turks but to resort to force of arms in order to fulfil their national aspirations.[133]

Fethi's cold reception, together with the 4 August speech of Lloyd

George, persuaded the Ankara government that the only option left was to prove themselves with the might of their military strength. Luck seemed to be on their side as well. This was related to a surprise movement of the Greek army in Anatolia. In July, as Ali Fethi started his unofficial visit to European capitals, the Greek prime minister made an unexpected decision and recalled two divisions from Asia Minor to join his army in Thrace. He then asked for permission from the Allies to enter Constantinople. Following the humiliating failure of Sakarya it seemed that nothing but the occupation of Constantinople could have restored the fortunes of the Greek army. The Allies, however, refused permission for the Greek troops march to Constantinople.[134] From Kemal's point of view the withdrawal of two Greek divisions from Asia Minor was very welcome news indeed, which diminished the number of Greek troops and made the troop numbers equal on both sides.

Mustafa Kemal ordered a general offensive on all fronts on 26 August 1922. The offensive started with a major Turkish attack on a 15-mile front south-west of Afyonkarahisar. The main part of the Turkish army started moving towards Smyrna. A secondary force started moving from Eskisehir in the direction of Bursa, the first substantial Ottoman capital. On the same day it became clear that the Greek forces were not going to hold for long. By the afternoon of the next day a Greek general retreat began. Within a few days the Greek army had totally lost morale and practically ceased to fight. General Tricoupis, the commander of the Greek armies, and another leading Greek general, Dionis, were both captured during the initial days of the battle. Leaderless and in disarray, the main Greek army retreated towards Smyrna with the Turks in hot pursuit.[135]

It is worth recalling the spirit of the Comintern at this point. Greek communists and the Bolshevik agents within the Greek army ranks carried out an extensive propaganda campaign which contributed to some extent to undermine the morale of the army. The Balkan Communist Federation (set up in Sofia in January 1920) had urged all communists to oppose the Greek campaign in Asia Minor.[136] This appeal had found its most effective response in the establishment of a special committee by the KKE in May 1921 – 'Central Committee of the Communist Soldiers of the Front'. The principal purpose of this committee was to undertake on the spot guidance of the anti-war work in the Greek army in Asia Minor.[137] KKE and the Balkan Federation sent professional revolutionaries and agents to the front to operate secretly within the Greek army. Their tasks were to shatter the morale of the army by propaganda and to operate closely with the 'Central Committee of the Communist Soldiers of the Front' in organising acts of sabotage and strikes. In Greece, too, the KKE launched an anti-war propaganda campaign and attempted to hinder the war effort by organising strikes and demonstrations against the Asia Minor campaign. Professor Novichev writes that 'the representative of the KKE declared during the Third Congress of the Communist International in July 1921 100,000 deserters or shirkers of military service avoided the Greek Army.'[138]

Just before the final offensive of the Turks in August 1922 the KKE launched an extensive anti-war campaign. Kousoulas writes that the communist militants of the KKE were '"instrumental" in spreading confusion and panic among the Greek troops during the critical days of August 1922 when the Turkish Army broke through the Greek lines.'[139] Although the effectiveness of the communist agitators could not be the main cause of the defeat it is fair to say that they contributed to a further disruption of the Greek lines when the Turkish army started its 'final attack'.

On 2 September the Greek government asked Britain to secure an armistice to permit the evacuation of Asia Minor by the Greek forces.[140] Curzon advised the Greeks not to panic or 'abandon good cards'.[141] After a week, on 9 September 1922 the Turkish army entered Smyrna and Bursa was captured the next day.[142] By the middle of September the Greek forces had evacuated all Anatolia. When the news about Turkish entry in Smyrna arrived in London, Lloyd George greeted the news by his widely quoted remark: 'Here is a bloody mess ...'[143]

The Chanak Crisis – War or Peace?

When the Greek army collapsed in Asia Minor, international attention was re-focused on the Straits. The region was protected by the British and their allies since the Mudros Armistice of October 1918. On 4 September, five days before the first Turkish troops entered Smyrna, Rumbold had informed London that it was the Allied positions in the Straits the Turkish forces would move on to after Smyrna. He recommended that the Gallipoli peninsula should be occupied as a last resort to secure British interests in the Straits.[144] A Turkish–British conflict seemed imminent and it soon became clear that Lloyd George's government would have been drawn into some action.

In the Cabinet on 7 September Lloyd George pronounced his intention of protecting British interests in Thrace and the Straits at whatever cost. The decisions which emerged from that meeting were as follows: the British forces would react by force if the Turks attempted to occupy the Gallipoli peninsula; the British naval force in the region would be strengthened; British forces themselves would occupy the Gallipoli peninsula; in the case of a Kemalist attack on Izmit the British forces would retreat. If the Turks attempted to occupy Constantinople, British naval and land forces would protect the city. The British priorities were thus set.[145]

The key strategic area was the so-called Neutral Zone on the Asiatic shore of the Straits. The purpose of the Neutral Zone was to protect both shores of the Dardanelles, the Sea of Marmara and the Bosphorus. The Neutral Zone did indeed consist of two zones in Anatolia linked by a narrow strip of coastline: opposite the Gallipoli peninsula the southern zone comprising the *vilayet* of Chanak and the town of the same name, and right opposite Constantinople across the Sea of Marmara the Izmit(Ismid) peninsula. There were 7,600 Allied troops in the Neutral Zone confronted by 6,000 Turkish troops near Izmit, 5,000 in the vicinity of Chanak, and an additional 40,000 coming from Smyrna.[146] As the British ground troops were outnumbered the major force against the Turks was the British fleet.[147]

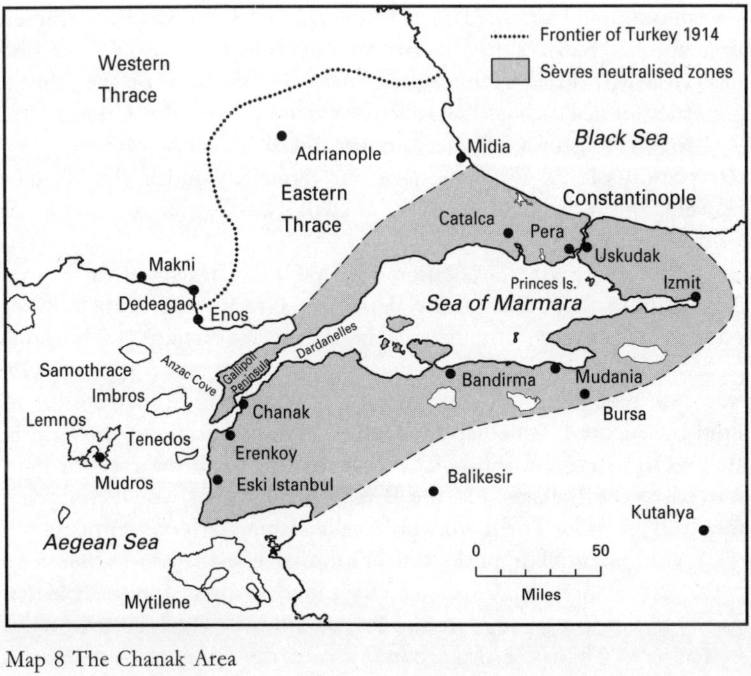

Map 8 The Chanak Area

The Turkish army was on the Asiatic shore and would have to be transported by water if it was to take either Gallipoli or Constantinople. The British fleet was in the Sea of Marmara and its position was very powerful. But the naval guns were limited in their effect when used against shore targets and the ships which carried them were vulnerable either to mines or shore batteries. The naval possession of the Sea of Marmara was less of an advantage than appeared at first sight.

The Gallipoli peninsula was held to be more important as it controlled the Dardanelles, the first passage on the way to the Black Sea. It seemed that Constantinople was the one more likely to be given back to the Turks because it was expected to appear as an absolute prerequisite of the Kemalists in any forthcoming negotiation. The priorities for defence were, therefore, judged to be Gallipoli, Constantinople, Chanak and Izmit peninsula in that order. It was thus agreed by the British that Gallipoli was first and foremost to their interests.[148]

On 12 September Mustafa Kemal gave his first official interview after the victory to Ward Price who had talked to him four years earlier in Constantinople. Price writes in his recollections that 'it was odd to remember that four years earlier it had been he who was in Constantinople had asked for an interview with me when he wanted to offer his services to the British government.'[149] In this second interview with Price, Kemal summarised the demands of the Ankara government following their decisive victory over the Greeks: 'We ask for Asia Minor, Thrace up to the river Meriç[Maritza], and Constantinople … We must have our capital … and if the Western Powers will not hand it over, I shall be obliged to march on Constantinople.'[150]

In the days after 9 September, as the Turkish army moved eastwards in the wake of the departing Greeks, the probability of open war between Britain and the Turks rose sharply. The zone itself, which was a creation of the Allies and the sultan's government, was not recognised by the Ankara government. The decision to hold the Neutral Zone had far reaching ramifications for the British. If the Turks passed through the Neutral Zone the war might have moved to the Balkans and the British government would find itself involved in more costly and politically damaging commitments.

A combination of pride and emotional imperialism, rather than *realpolitik*, made the British leaders eager to fight for the Neutral Zone. It was the prestige of the British Empire which seemed to be in danger at Chanak. Lloyd George was of the opinion that if Britain lost control of the Straits it would have an extremely negative impact upon the wider interests of the Empire.[151] Churchill provided further support to this attitude by saying that ,

> The line of deep water separating Asia from Europe was a line of great significance, and we must make that line secure by every means within our power. If the Turks take Gallipoli and Constantinople, we should have lost the whole fruit of our victory.[152]

The whole affair escalated to the point of a diplomatic crisis, known as the 'Chanak crisis'. International attention focused on the narrow strip of land and the shores of the Straits and a sudden likelihood of a new wave of hostilities in the aftermath of the Greco–Turkish war threatened international politics. There was now the possibility of confrontation between Turkish troops and the British forces which had held the area since the end of the war. The Chanak crisis also threatened to develop into a serious conflict between Britain and Soviet Russia. The Turkish lands had for the last three years been considered a most sensitive region and potentially the main source of any possible clash between the two powers. Their respective roles in the Anatolian conflict, Britain as the power hostile to the Turkish nationalists with the Bolshevik government as the Turks' main ally, had been established earlier.

Things had changed with the March 1921 Trade Agreement which conceded to the young Soviet republic partial recognition by the British government. There was a growing tendency on both sides to manage their relations without committing their military forces to a regional war. In the Anatolian conflict, London declared its neutrality in March 1921 and carefully avoided association with the Greeks' fatal campaign in Asia Minor. Moscow also avoided agitating Turkish nationalists and lessened the risk of attracting a British protest that might have weakened the international relations of the Soviet state.

In this way, until the recent crisis, both Britain and Soviet Russia had avoided direct involvement in the Anatolian conflict. The state of British–Soviet relations during the period of March 1921–September 1922 represents a regional system in which certain restraints and norms were present and adhered to. Both parties were anxious to prevent the further spread of the influence of the other. But they were also careful not to push further interests which might escalate a regional war between Britain and Soviet Russia. In late 1922, as a result of the decisive defeat of the Greek army, Britain and Soviet Russia were drawn into a likely confrontation neither had actually willed. When the Turkish troops started to move against the

British positions at Chanak, British policy-makers in London had to reconsider not only the capabilities and intentions of the Turks but also the role of the Bolsheviks in this recent confrontation.

London was concerned that the Bolsheviks would try everything to influence the situation to their benefit. The reply of Chicherin at Genoa to Lloyd George's demand to adopt a policy of neutrality in the Anatolian conflict was still fresh in their minds. 'At Genoa M. Chicherin had expressly stated that in the event of war with Turkey, Russia would not be bound by the pact of non-aggression'.[153] Chicherin's words were interpreted as Soviet Russia's determination to intervene in Turkey against the British forces.

On 16 September when the first Turkish advance parties were approaching the borders of the Neutral Zone from the south, Karakhan sent a note to Curzon that the solution to the Straits problem could only be solved according to the Turco–Soviet treaty of 16 March 1921. This was said to be 'the only international agreement since the war on the question of the Straits which is in force'. Karakhan in his note accused the Allied powers as being responsible 'for the disturbance of the peace in the Black Sea and for the heavy damage which has thereby been inflicted upon the economic life and the commercial activities of the Straits'. He further stated that:

> Russia cannot consent to the Straits being opened to the battleships of any country and in particular that Great Britain, with the consent of her allies, should have control of the Straits without the consent and against the wishes of the Powers who have vital interests in the Black Sea and who should have the right of decision as to the fate of the Straits … Russia, Turkey, the Ukraine, and Georgia, to whom belongs nearly the whole of the Black Sea, cannot admit the right of any other government to interfere in the settlement of the question of the Straits, and will maintain the point of view above set out even if the contrary point of view is backed by military or naval superiority.[154]

Crisis deepened with an official press release from Ankara which indicated that the Turks were in favour of the Soviet military involvement.[155] The Kemalist government officially requested Soviet action against the Balkan states if they intervened in the Chanak conflict alongside Britain. To this request Aralov replied, referring to a message from Karakhan, that the Turkish army should endeavour to liberate Constantinople and the Straits.[156]

Various reports coming from the region fuelled the British government's suspicions of a possible Soviet involvement. According to an intelligence communication coming from Tiflis, the Soviet government was reported to be discussing wide-ranging military and diplomatic measures in support of Ankara. The communication reports that military preparations and reinforcements would be sent to the Balkans and Caucasus. There was also information about their transport route across the sea: with the assistance of the Bolsheviks the Kemalists intended to transport troops by motor lighters from Zonguldak to the Midia district and via Constantinople to the Silivri area.[157]

On 24 September the foreign ministers of Great Britain, France, Italy, Yugoslavia, Bulgaria, Romania, and Greece, and the prime minister of Egypt received a long telegram from Moscow. It protested against the British designs in regard to the future of Constantinople and the Straits. The Soviet government maintained that the international statute for the Straits should be drawn up by the Black Sea powers parallel to Article 5 of the Turkish–Soviet treaty signed in Moscow on 16 March 1921. Moscow further warned that:

> no decision on the Straits taken without Russia will be final and enduring. It will merely sow the seeds of new conflicts. The freedom of the Straits which Great Britain has in mind signifies only the desire of a strong naval power to control a route vitally necessary to other states in order thereby to keep them under a constant threat. This threat is directed in the first place against Russia and Turkey. Great Britain is dispatching military forces to the Near East, and is trying to drag France and Italy, as well as Yugoslavia and Romania, into the war with Turkey.[158]

Moscow's unveiled threat made its position clear to the West, particularly to the British. The key question then was how serious the Bolsheviks were in their intentions, and how far they were determined to go. Some Soviet leaders would probably have been more than pleased to see Britain enter into an open war with the Turks in the region. It would have increased the dependency of the Ankara government upon Soviet assistance and strengthened their 'special relationship'. Aralov made reference to such an expectation in his memoirs.[159] However, the Soviet government as a whole would probably not be willing to jeopardise their international position as a result of direct involvement. It could be too risky to enter into any direct clash with the British at Chanak, only a few hundred miles from their sensitive Black Sea borders.

Consequently when the Turks officially demanded Moscow's direct involvement, the Soviet government advised Ankara to communicate directly with the Ukrainian government at Kharkov and to ask the

Ukrainians to intervene. The Ankara government thereupon sent a long communication to the Ukrainian government with reference to their separate treaty agreement with the Ukraine of January 1922. The Ankara note terminated with a request for a response as to whether the Ukraine would act in accordance with the provisions of that treaty.[160] Perhaps not unexpectedly, no response came from Kharkov.

It is not at all surprising to conclude that of all the major participants in the Chanak crisis Moscow's behaviour was the most restrained and cautious. Moscow carefully restricted itself largely to diplomatic demarches, political contacts, and the signalling of intentions through policy statements and the media. Soviet representatives merely made demands. They placed the burden of initiating the use of force on the British.

The reluctance of the Bolsheviks to be involved directly in the Chanak conflict did not go unnoticed by the British. Even as early as 9 September, the day when the Turkish troops entered Smyrna, there was some indication of this. When the news about an agreement between the Soviet government and its most important British claimant, the Russo–Asiatic Consolidated Company Ltd., arrived in London, there appeared the important opinion that the Bolsheviks still valued their relations with the West. It was the essence of the agreement which caused that conviction.[161] The British company, represented by John Leslie Urquhart, was making claims against Moscow amounting to 56 million pounds. This represented about one-third of the total British claims.[162]

Another indication of Moscow's cautious approach arrived about 10 days later when Curzon received a telegram from Mr Hodgson, head of the British Commercial Mission and the official British agent in Moscow.[163] Hodgson summarised his conversation with Radek, who gave an interview as political adviser on foreign affairs in the Politburo of the RCP(B). Radek was convinced 'that the time had now come when British and Russian views on the Near East question could be harmonised. Russia could be opposed to the Turks gaining a footing in Europe or controlling the Dardanelles and would accept the solution which would internationalise the latter while respecting Turkish sentiment.'[164]

Both the Krasin–Urquhart agreement and the conversation with Radek were positive signs for the British, but obviously not enough to eliminate suspicions about a possible Soviet involvement. An imminent direct Soviet involvement might be a bluff, but how could it be possible to establish this with some degree of confidence?

After all, Britain could not really know, without testing, whether the Bolsheviks would stand firm with the Turks or not.

Two days after the information arrived from Moscow about Radek–Hodgson meeting a contradictory SIS intelligence report was received in London. According to it both Karakhan (in Moscow) and Aralov (in Ankara) promised full military support to the Turks.[165] Another came on 25 September this time exploring the different policies of leading Bolsheviks:

> Trotsky and Bukharin are advocating military action ... but Chicherin is opposing this on the grounds of the dangerous internal complications ... the loss of advantages already secured in Europe, the probability of a blockade of Russian ports, also the difficulties which would be involved in maintaining the numerous resultant fronts.[166]

These contradictory reports made it all the more difficult to judge the real intention of the Bolsheviks. Despite the existence of some positive signs about the Soviets' intention for non-involvement Moscow's position in the conflict remained a question mark for the British government. During the whole Chanak affair Britain had kept an anxious eye on the Soviets.

On the other hand the Turks seized the opportunity and enjoyed their tactical advantage in the face of British uncertainty. By late September it became evident that Mustafa Kemal's show of force was building up to substantial proportions. On 25 September a Turkish cavalry unit with machine guns entered the Neutral Zone at Erenköy, about 20 miles south of Chanak.[167] No opposition was offered by the British.[168] When demanded by the British to clear the Neutral Zone, Mustafa Kemal stated that: 'up to the present I have not been informed of any neutral zone having been established between the governments concerned and the government of the Grand National Assembly. The movements of our cavalry and other units consist in following up the retreat of the hostile Greek Army.'[169]

Meanwhile another Turkish force entered the Neutral Zone from the direction of Biga, about 50 miles east of Chanak.[170] The influx of the Turkish forces continued from every direction and by 29 September they were in close contact with British troops. They had reached the barbed wire of British position and were making grimaces across it. A few shots were fired, and two British soldiers were killed.[171]

On the same day as the Neutral Zone was crossed by the Turkish troops, the Allies agreed to call for an armistice. In the name of the Allied powers M. Franklin-Bouillon went to Smyrna and met Mustafa

Kemal on 28 September. During this meeting a memorandum signed by the Allied foreign ministers asking for a conference at Mudania or Izmit was presented to Kemal.[172]

Following this meeting with Mustafa Kemal, Franklin-Bouillon came to the conclusion that there was no real likelihood of Soviet–Turkish military co-operation in the Straits and the Turks were willing to accept the Allied offer for a conference.[173] A similar conviction was expressed the next day by the British G.H.Q.(Constantinople).[174] Two days later, on 1 October, Lloyd George was talking to the Romanian representative with authority confirming that 'Soviet Russia was still anxious for friendly relations with Great Britain …' and that the British were able 'to exercise a restraining influence upon Russia and to persuade her to remain neutral.'[175]

The crisis at Chanak duly came to an end when Mustafa Kemal accepted the Allied offer for having a conference of armistice at the end of September 1922. On 3 October the conference was opened at Mudania, a small sleepy port north of Bursa on the Sea of Marmara, and on 11 October the Mudania Conference concluded an armistice.[176] The open war in the Anatolian conflict was finally brought to an end. Given the threatening escalation of the events surrounding the Chanak affair it was rather a peaceful end.

6

Settlement
September 1922–August 1923

> Turkey knows neither gratitude nor resentment; her only consistent political emotion is fear. For a century she was afraid of Russia and made friends with us; when we in our turn made friends with Russia Turkey sought refuge with Germany; now that she is afraid of us ... It is probable that in a few years she will be afraid again of Russia and will turn again to Great Britain. [H. Nicolson, 3 October 1922][1]

> When the English pursued the Near East and the Straits, when the peoples of the East came to be involved in our relationship, as they were at Lausanne, the English and ourselves found each other in opposite camps. [C. Rakovski, December 1922][2]

> Being a *mariage de convenance*, not founded on spiritual affinity, it could retain no hold on the parties when the particular international situation that led to it had passed away. The Turks have not been looking to Russia for anything but munitions, diplomatic support and gold; they have not been learning her language, studying in her universities, buying her goods, or imbibing her culture. The West, and not Russia, remains their ideal; the Western comity of nations, not the Third International, is the society into which they are seeking an entry; and now that the supply of material benefits from Russia is running low, while a settlement with the Western Powers is slowly but steadily coming nearer, they are beginning to remember their relations with Russia in the past and to look forward with increasing uneasiness to the future. [Arnold J. Toynbee, June 1923][3]

The Call for Lausanne

When the likelihood of a Turkish–British war was successfully driven away it became possible once more to turn an eye to issues other than the war itself. The major issue to surface was the nature of the long-waited Turkish settlement. After the signature of the armistice at Mudania preparations started for a peace conference.[4] The British were convinced that in order to show 'the Turks of the good faith of the allies' the conference should open as soon as possible – 'in the week beginning November 7th, if not, at latest on November 14th.'[5] This was considered particularly important to break up the Turco–Soviet alliance and to prove to the Turks that, in Toynbee's words, 'the Western comity of nations, not the Third International, is the society' which the new Turkey should look at. By showing 'good faith' London aimed to remove Turkey out of Soviet Russia's sphere of influence.

After arguing over Prinkipo, Istanbul and even Usküdar, Lausanne was selected as the meeting place for a number of reasons. First of all it was in a neutral and disinterested country – Switzerland. The town was also on the direct line of the Orient Express to Constantinople. It had excellent hotel accommodation and a good climate, even in winter, which would relax the tired minds of diplomatic envoys in between sessions of long and exhausting negotiations. Finally, because of its central position in Europe, it would be possible for the foreign delegates to go to and fro with comparative ease.[6]

The first important problem – invitations – emerged even before the opening of the conference. Invitations were issued on 20 September to Britain, France, Italy, Romania, Yugoslavia, Japan, Greece, Turkey, and the United States to meet at Lausanne.[7] This caused a stream of protest from the Soviet government. Moscow had been expecting an international conference, but with the direct participation of Soviet Russia, the Caucasian Soviet republics and the Ukraine.[8] Also, in anticipation of an important role in the forthcoming settlement it was looking forward to joining the conference as a major power. The Soviet government's position was presented in a series of articles and official statements in the official newspaper, Pravda, under the headlines such as, 'Russia and the Near East',[9] 'The Problem of Constantinople and the Straits',[10] 'Turkey Will Not Go to the Conference Without Soviet Russia',[11] and 'Participation of the Soviet Russia to the Conference'.[12]

Furthermore, during the Chanak crisis the Soviet government had officially demanded participation in an international conference to resolve the Turkish question.[13] This position was supported by the Ankara government in its reply of 4 October to the Allied invitation for the peace conference. The Ankara note stated:

> Only important question of this kind is the future control of the Straits. We cannot but express surprise that Russia, the Ukraine and Georgia, who are deeply interested in this, should not have been invited, as the participation of these would only help to make the settlement more lasting,

> and to avoid all future cause for conflict we definitely propose that they should be invited ... and we hope that the invitation will be addressed to them before the conference.[14]

Ankara naturally expected the Bolsheviks, the Turks' only ally throughout the nationalist campaign, to be invited. It would ensure that the Turkish delegation at Lausanne would not be isolated completely vis-à-vis the Allied Powers. The Turks did not want to be faced with 'a united front' of 'the European nations' alone.[15] It might also offer the opportunity to play one power against the other to gain some trade offs.

On 19 October Moscow sent a further note to the British government expressing astonishment that their previous note had not been answered and stating that the attempt to confront the Soviet government with the *fait accompli* of decisions arrived at without its participation would compel them to refuse to recognise those decisions.[16] Moscow had been warned before that it was indeed the intention of the British to leave them out. Berzin, the deputy official representative of the Soviet government in London, had reported to Moscow earlier that he had gained the impression that the British would not invite the Soviet government to a possible forthcoming international conference.[17] Aralov, writing from Ankara in early October, concluded that it was unlikely that they would be invited.[18]

The British, French, and Italian governments, on the other hand, confirmed that they are 'invited to the conference, which is to assemble at Lausanne to conclude the state of war in the East by the conclusion of a treaty of peace with Turkey, the states which are actually not in a state of definite peace with the latter power.' Since Soviet Russia, the Ukraine and Georgia had regulated their affairs with Turkey by the treaties of Moscow, Ankara, and Kars respectively, the Allied governments maintained that these three states were not party to the discussions to be raised at Lausanne.[19]

The situation was rescued by a compromise before it became stalemate. The Soviet delegations could be admitted to the Lausanne conference in order to participate in the discussion of the question of the Straits only. A formal invitation was then handed to the Soviet government on 27 October 1922.[20]

On the same day Lenin gave an interview to the correspondent of the *Observer* and *Manchester Guardian* in Moscow. Describing the British reluctance to admit the Soviet government to fully participate in the conference as an 'act of injustice', Lenin issued a warning that it was not possible under these circumstances to ratify the Krasin–Urquhart Agreement. He asserted that it was 'a direct expression … of … the sentiment of the entire mass of the workers and peasants.'[21]

It is interesting to note in this context that the earlier conviction which had emerged with the news about the Krasin–Urquhart agreement in early September suddenly appeared to be too optimistic. Maybe it was true that the Bolsheviks still valued their relations with the West, but they were not in favour of a unilateral compromise. They wanted to get something in return. A central role in the Turkish settlement seemed to the Bolsheviks a stake worthwhile to go for. Moscow used non-ratification of the Krasin–Urquhart agreement for distinctly political ends. During the following months the Soviet leaders consistently argued that non-ratification directly resulted from Britain's hostility towards Soviet Russia on the Turkish question, and that only when the attitude of the British government changed could there be any hope for an improvement of Anglo–Soviet commercial relations.[22]

The Fourth Congress of the Comintern, which was held in Moscow between the receipt of the invitation to Lausanne and the opening of the conference, clearly showed how the Bolshevik leadership was sensitive about the issue of the Turkish settlement.[23] The forthcoming peace conference was one of the main points in the agenda. Two meetings were devoted to the issue and the forthcoming peace conference was discussed by the delegates in detail. During these discussions the importance of Turco–Soviet solidarity at this critical junction was emphasised, and the imperialist policies of the British and their allies were criticised severely.[24]

A Strained Friendship

The peace conference was not the only item related to Turkey in the agenda of the Fourth Congress of the Comintern. On 19 November at the 18th session, Orhan (Sadrettin Celal), the leading delegate of the Communist Party of Turkey, complained that the Ankara government had started a ruthless campaign of repression against the communists. Celal asserted that the recent anti-communist campaign of the Kemalist government was directly related to a serious change in the orientation of the Turkish nationalist policy on the eve of the peace settlement. He concluded that having won the war against the Greek army, the Kemalists now were anxious to prove to the West that they had nothing in common with Soviet Russia and its ideology, communism.[25]

Indeed, during the short period of less than three months from the Kemalist victory on 30 August to the opening of the Lausanne conference on 20 November 1922, left-wing and communist activities in Anatolia had been violently suppressed and virtually brought to an end. During this period Aralov was in continual conflict with the Ankara authorities over the persecution of Turkish communists. The Kemalists refused to allow the communists to hold public meetings and ignored Aralov's protests. By mid-November more than 200 communists had been arrested, a number of labour organisations had been closed down and their leaders were put in prison.[26]

Moscow was following the rising anti-communist tide in Turkey with an increasing unease. Soon after the Kemalist victory in western Asia Minor, the secretary of the Comintern's Eastern Bureau issued a statement describing the changing political conditions in Turkey. It was stated that the Kemalist administration had initiated a wide spread anti-communist witch-hunt. The Comintern was also concerned about the imminent take-over of Constantinople by the Turkish forces as they expected Kemalist police to put an end to all communist and left-wing activities there. The Constantinople communists had flourished under the Allied occupation as a result of the inefficiency of the weak Ottoman government and the ignorance, or perhaps indifference, of the Allied police of the language and local circumstances. These groupings were now to face a more experienced and determined police force of the Ankara government.[27] As a result of disturbing news from Turkey Zinoviev, the head of the Comintern, on 14 November sent a letter to the leading Bolsheviks (Lenin, Trotsky, Radek and Bukharin) urging them for action to stop the ongoing persecution of the Turkish communists.[28]

Although disturbing these incidents were not unexpected for the Bolshevik leadership. The strong Western orientation of the Kemalist movement and well-determined anti-communism of the nationalist leadership had been well-known in Moscow since the winter of 1920–21, when the Ankara government had undertaken another severe campaign against the communists. Aralov recollects in his memoirs how Lenin had briefed him about Mustafa Kemal before leaving for his post in Ankara in late 1920. In Lenin's opinion Mustafa Kemal was 'an opportunist', not a revolutionary, and he expected Kemal to be bought by the highest bidder. There was no illusion about the Turkish nationalists as far as the Bolshevik leaders were concerned.[29] Nevertheless the Bolshevik strategy called for the exploitation of every possible opportunity. This pragmatism was later described in a book written by Bagirov who ascertained that the Bolsheviks knew that as long as the military emergency in Anatolia continued Ankara had to rely on the Soviet military and diplomatic assistance, and they contented themselves with that much. They were determined to help the Kemalist movement to counter-balance the influence of the West in the Near East.[30]

When the Turks won the military victory in the Anatolian struggle it became more than apparent that the 'life-support' urgency of Soviet help had lost its appeal for Ankara. As the Greeks were out of sight Ankara was keen to strike a deal with the West.

Following the final Turkish victory over the Greeks in Anatolia, on 30 August 1922 the Soviet mission in the nationalist capital sent a number of reports to Moscow where an apparent shift in Turkish policy was identified. They warned Moscow against a probable isolation of the Soviet government at Lausanne. The Soviet mission in Ankara also drew attention to a number of unresolved mutual problems. The problem of the Molokan minority living under Turkish rule in the eastern Anatolian provinces was still unsettled. And there was always the possibility of an anti–Soviet Turkic–Muslim movement in Central Asia and Caucasia where sizeable Turkic-speaking communities lived.[31]

Clear evidence of Ankara's distancing itself surfaced when Moscow demanded the right to open a consulate in Smyrna. Just after Smyrna was put under Turkish control in early September Aralov had intended to visit the city and to open a consulate there. But despite his insistence the Turkish authorities did not invite him to the city. Neither did the Soviet representative receive an official response from the Turks regarding Moscow's demand to open a consulate in Smyrna. On 4 November Aralov met Ismet Pasha and raised the issue, emphasising the fact that almost all the Western powers had representatives in Smyrna. Avoiding a clear response Ismet stated that they would consider the matter in the near future.[32]

Despite all the negative reports from Turkey, the Soviet leaders held the view that it was not the end of the game. They still believed that a satisfactory settlement acceptable by all sides would need time to come and that their active interference might determine the outcome of the final settlement. It did not seem politically wise to offend and even upset Ankara. Moscow preferred not to issue a strong official protest over the anti-communist campaign and did not mention again the proposed consulate in Smyrna. Furthermore, Moscow advised its representatives in Turkey to adopt a more conciliatory approach and to keep relations at a stable level with the Ankara government. The mission was told not to become directly involved in the activities of local communists. Moscow took this cautious approach one step further by blaming the 'careless adventurism' of the Turkish communists and accused them of causing unnecessary anti-Soviet sentiments and mistrust in Ankara.[33]

Aralov categorically denied that he was involved in any scheme 'behind the back' of the Kemalist government. Fully aware of the anti-communist policy of the Ankara government, Aralov still believed that even the smallest concession from the Turkish government would be of great use for the long-term Soviet interests in the region. He reiterates in his memoirs that the geographical location of Turkey, ever important for the protection of the sensitive southern borders of the Soviet state, gave a special status to the Bolsheviks' relations with Ankara and they simply did not want to offend the Turks.[34]

With similar considerations the Executive Committee of the Comintern issued a statement to 'the male and female workers of Turkey' two weeks before the Mudania Armistice, calling them to continue their support for the 'bourgeois-nationalist leadership' of the independence movement in the struggle against Western imperialism led by Britain.[35] This policy was repeated during the debates on the Eastern Question at the Fourth Congress of the Comintern. Although the Kemalist administration was criticised for its campaign against the communists, the overriding importance of the national independence struggle against Western imperialism was emphasised. Turkish communists were asked to give unconditional support to the Ankara government in order to secure a favourable peace settlement.[36]

Negotiations

While the Soviet government was trying to restrain the Turkish communists and to maintain the fragile balance, the British were getting ready for the negotiations. During the three year period after the end of the First World War the British thought the Turkish question could be resolved by the military operations of the Greek army. The policy was undermined after the Greek army's first unfortunate campaign in the spring months of 1921, but the British government had been reluctant to revise its policy radically. Despite all attempts by the Kemalists to convince the British that Turkish–British co-operation would better guarantee the safety of British interests in the region, Lloyd George's government had kept its pro-Greek policy intact with the hope for a decisive Greek victory in Anatolia until late August 1922. Once it was irrefutably clear that the Greek adventure failed in Turkey, the anti-Turkish stand of Lloyd George's government was dealt a decisive blow.

In late October Lloyd George fell from power. He and his political associates had planned a general election for 28 October to forestall the Conservative Party Conference scheduled for 15 November. But he and his friends had been in turn forestalled by what later became the 1922 Committee meeting at the Carlton Club on 19 October. On 23 October, four days after Lloyd George's fall, Andrew Bonar Law took on the task of forming a new government and organising the new elections that would take place in mid-November. In the new government Curzon retained his office as foreign secretary and he took charge of the negotiations at Lausanne.[37]

It would be too simplistic to claim that Lloyd George fell from office solely because of the failure of his Greco–Turkish policy. There were far too many trouble spots, both domestic and international. Ireland, increasing difficulties in Egypt, the burden of unemployment, the impact of the coal strike, and the state of affairs with Soviet Russia all deserve to be given serious consideration. Not one of them on its own could account for the historic 1922 Committee decision, nor is it easy to order their significance. In the minds of Tory backbenchers, these mounting pressures all combined with the bankruptcy of the Anatolian enterprise to justify their action. As for the Turkish settlement, the change meant that Curzon's task at Lausanne would be easier to the extent that he 'need not fear his prime minister was plotting behind his back'.[38]

The change in Britain's Near Eastern policy was confirmed in a memorandum by the General Staff on the same day that Lloyd George fell from the power. The memorandum reads:

> we must recognise the re-establishment of Turkish power, it will surely be to our advantage to do everything possible to give the new Turkey the chance of restoring order in her own house and defending her territory. The Turks are numerically a small nation, and, surrounded by potential enemies as they are ... Fate, however, has located them in one of the most coveted areas in the world, and if we are to have peace in the Near East they must be strong enough to defend it. They can attack no vital points in the British Empire (neither Iraq nor Palestine can be considered as such), and therefore, from the point of view of the General Staff, so long as our relations with them are friendly, it is to the advantage of HMG to strengthen them in a military sense rather than the reverse.[39]

Thus the Lausanne conference started with a new British government, a friendly approach from Ankara keen to impress the British and the Bolsheviks trailing behind to keep up appearances. The conference continued more than eight months – starting on 20 November 1922 and ending on 24 July 1923 with the signature of the peace treaty. The negotiations were adjourned on 4 February 1923 and re-opened on 23 April 1923. During the Lausanne talks the British aim was to facilitate the emergence of a Turkish state that would act as an effective buffer next to the Soviet-controlled lands. It was considered necessary and legitimate to do everything to dissolve the Soviet–Turkish alignment. To this end Curzon, the head of the British delegation, aimed to isolate the Bolsheviks by achieving a satisfactory settlement with the Turks.[40]

With the opening of the conference the competition between Britain and Russia, the two powers chiefly concerned in the region from at least 1830s onwards, was explicit. And as in the previous century, the question of the Straits once again became a major bone of contention between two countries and proved to be a more serious problem between them than it was between the Turks and the Allies. It was discussed early in December in the Territorial and Military Commission.[41] The conference was faced with two basic theses concerning the problem of the Turkish Straits. First was the British position which was designed to maintain the principle of transit and navigation of the Straits under an international regime. This, in essence, would preserve the dominant influence of British sea power in the region. The second was the Soviet project which provided for Turkish sovereignty but would close the Straits to foreign warships and convert the Black Sea into a Soviet *mare clausum*.

Both the British and the Soviet positions on the Straits question appeared to be the reverse of the two countries' respective positions before the First World War. Up to 1915 British policy had been directed towards keeping the Russian Black Sea fleet out of the Mediterranean (the closing of the Straits) rather than to enabling the British Mediterranean fleet to enter the Black Sea (the freedom of the Straits). The tendency throughout the nineteenth century to press for the closing rather than the opening of the Straits could be explained by an important consideration. So long as Russia coveted the Straits, and Britain was opposed to the realisation of this ambition, it was inevitable that the Ottoman sultan would look to Britain and France for protection in the event of a Russian attack. Under the convention of 1841 the sultan retained the right, if Turkey were at war, to allow foreign warships to pass through the Straits. It thus became almost an axiom during the nineteenth century that the sultan would open the Dardanelles, and even the Bosphorus, to the British fleet in the event of any Russian attack. The policy of the closing of the Straits was inspired by the assumption that in the last resort the Ottoman ruler, as in the Crimean war, would let the British navy in.

The Turkish position had changed fundamentally during the First World War. The Ottoman Empire on 12 August 1914, by allowing the German battle cruiser 'Goeben' and the light cruiser 'Breslau' to pass up the Dardanelles and to take refuge in the Golden Horn, violated the Straits Convention.[42] The subsequent closing of the Straits, even to commerce, and the ensuing Gallipoli campaign convinced the British government that the geographical position of the Ottoman Empire placed in its hands an immense opportunity for obstruction. Britain concluded that one of the objects of the war should be to remove this vital corridor from Ottoman control. With the defeat of the Ottoman Empire the new doctrine of the freedom of the Straits reflected a determination on the part of the British government to ensure that never again would the sultan to be able to bar the passage of the British fleet into the Black Sea. There was another, more pragmatic reason for this shift in the British position. At the time of the Lausanne Conference the Soviet navy was almost non-existent. Most of the Russian vessels in the Black Sea had been taken over by Wrangel and then by the French. For the British and their allies, freedom to enter the Black Sea meant the possibility of keeping up a certain pressure upon the southern borders of the Soviet Union.[43]

On the other hand, the Russian rulers throughout the nineteenth century had thought in terms of expansion, of access to the Mediterranean. Now, however, Moscow wished only to be left in peace at its home base and to be protected against possible attacks on its southern coast. In a newspaper interview given early in October 1922 Lenin described the programme of the Soviet government as 'the closing of the Straits to all warships in times of peace and war.' This was considered essential for the security of the Soviet state. On the question of the Straits Lenin preferred a minimalist approach: 'We should like to see a minimum of general assurances, solemn promises and grandiloquent formulas, and the greatest possible number of the simplest and most obvious decisions and measures that would certainly lead to peace.'[44] During the negotiations at Lausanne Chicherin[45] demanded that commercial navigation and non-military maritime communication through the Straits should be absolutely free.[46] However in order to maintain permanent peace in the Black Sea, he argued that the Straits should be permanently closed to all men-of-war and aircraft of all countries except those of Turkey.[47] Curzon noted sarcastically that Chicherin found himself to be 'more Turkish than the Turks'.[48]

The Turkish delegation, for its own part, informed Curzon in private that they were ready to give their consent to the Allied proposals provided that the British would support the territorial demands of the Turks.[49] The Turkish government favoured the British terms of settlement because they would guarantee security for Turkey against Soviet power in the Black Sea. Turks did not particularly welcome the prospect of finding themselves face to face with the Soviet navy in the Black Sea while warships of all other nations were excluded. If the Black Sea were to become a Soviet lake, the Soviet navy would be able to control both the Black Sea and the Straits. The lesson Turkey learned from history was that in order to maintain a balance of power in the Black Sea, the presence of other European warships was essential. The Turks, therefore, were ready to back up the British proposal on the Black Sea against the Soviets.[50]

When Vorovsky arrived in Lausanne, as the first member of the Soviet delegation, he met the head of the Turkish delegation,[51] Ismet Pasha,[52] in order to secure a common policy against the British proposals. *Pravda*, reporting about this meeting wrote that 'the meeting of Vorovsky with Ismet Pasha was regarded in London as an evidence of strong Russo–Turkish understanding.'[53] The Turkish representative, however, did not agree. In his own report to Ankara on 2 December 1922 Ismet Pasha was convinced that to support the Soviet point would isolate Turkey and threaten a settlement with the Western Powers.[54]

When it became apparent that the Turks were ready to accept the British proposal with some minor modifications, the Soviet delegation found itself at the conference both isolated and deprived of the main argument on which the Soviet position was to be based. The Soviet government accused the Turks of a capitulation which would put Turkey under the political yoke of British imperialism. The Turkish position was, moreover, untenable as it was claimed to be in breach with the Article IV of the 1921 Soviet–Turkish Treaty of Friendship. Ismet Pasha was officially warned by the Soviet delegation that 'the admission of the war vessels through the Straits would be a hostile act against the Black Sea states'.[55]

The Soviets failed in their efforts to persuade the Turkish delegation and were thus isolated. After further discussions, the conference resolved that with regard to the Straits freedom of passage would be established for merchant ships in peace or war.

Furthermore, freedom of passage would also be granted to all warships in time of peace. However, it was resolved that:

> The maximum force which any one power may send through the Straits into the Black Sea is not to be greater than that of the most powerful fleet of the littoral Powers of the Black Sea existing in that sea at the time of passage; but with the proviso that the Powers reserve to themselves the right to send into the Black Sea, at all times under all circumstances, a force of not more than three ships, of which no individual ship can exceed 10,000 tons.[56]

Other provisions decided on were that an International Straits Commission would be established and that the Dardanelles and Bosphorus should be demilitarised. The draft convention on the Straits was approved by the conference on 1 February 1923.[57] According to the Straits Convention of Lausanne, which was finally signed on the same day with the Treaty of Lausanne on 24 July 1923, Turkey obtained a right to station a garrison at Constantinople and on the Gallipoli Peninsula. Among other things, Article 1 of the convention stated: 'The High Contracting Parties agree to recognise and declare the principle of freedom of transit and navigation by sea and air in the Straits of the Dardanelles, the Sea of Marmara and the Bosphorus, hereinafter comprised under the general term of the "Straits."'[58]

The demilitarised zones of the Straits included the Gallipoli Peninsula, and the European and Asiatic shores, to a length of some 75 miles. The zones ran the entire length of the Bosphorus to a depth of 9 miles. No permanent fortifications, artillery organisation, submarine engine of war, military air base, or naval base was to be permitted in the demilitarised zones, subject to the exception that a garrison of 12,000 men could be maintained at Constantinople and a naval base and arsenal could be constructed there. To implement the principle of freedom of the Straits an International Commission of the Straits, as already observed, was established at Constantinople composed of one representative each from the following countries: Turkey, France, Great Britain, Italy, Japan, Bulgaria, Romania, Soviet Russia and Yugoslavia.

Turkey had desired an individual and collective security guarantee for the zone of the Straits, a proposition which the Allied Powers rejected. Instead, under Article 18, they stipulated:

> Should the freedom of navigation of the Straits or the security of the demilitarised zones be imperilled by a violation of the provisions relating to freedom of passage, or by a surprise attack or some act of war or

> threat of war, the high contracting parties, and in any case, France, Great Britain, Italy and Japan, acting in conjunction, will meet such violation, attack or other act of war or threat of war, by all the means that the Council of the League of Nations may decide for this purpose.[59]

The Soviet government at first denounced the Lausanne Convention. Chicherin replied at length on behalf of the Soviet government, recapitulating their objections to the Straits Convention and manner in which it was drafted. Rakovski proclaimed that 'We shall not give our signature to the transformation of the Black Sea into an English lake'.[60] To show, however, their peaceful intentions and without withdrawing objections to the convention or 'to the so-called League of Nations', the Soviet delegation decided to collaborate with the 'Entente Powers'. It was announced that Moscow would send its representative at Rome, M. Jordanski, to Constantinople to sign the Straits Convention there'.[61] Monsieur Jordanski signed the Straits Convention in Rome, not in Constantinople, on 14 August 1923. But the convention was never ratified by the Soviet government.[62]

The Crisis in Anglo–Soviet Relations

The participation of the Bolsheviks in the discussions at Lausanne remained limited to the question of the Straits, at least at the official level. After February 1923 the Bolsheviks followed the discussions as observers. An unfortunate incident further offended the Soviet government and deepened their bitter feelings over the whole Lausanne episode. Vorovsky, the Soviet representative in Italy, had been sent to Lausanne as head of the unofficial Soviet delegation to observe the rest of the negotiations between the Allies and the Turkish government. He was shot dead in the dining room of his Lausanne hotel, Hotel Cecil, on 10 May 1923 by a Swiss citizen, M. Conradi,[63] who had formerly been in the Russian army.[64] Two of Vorovski's assistants, Ahrens and Divilkovski, were wounded in the same attack. Following the death of Vorovsky Moscow accused the Swiss government, as well as the Allied powers who had organised the conference, of failing to take 'the least precautions for the protection of the Russian–Ukrainian–Georgian delegate and his colleagues'.[65]

Vorovsky's assassination represented the nadir in the fast deterioration of British–Soviet relations during the Lausanne conference.[66] A series of events which contributed to the existing mistrust and hostility between the two sides was brought to a head by this politically motivated murder. Starting with the debate about conference invitations and re-visiting the familiar Anglo–Russian rivalry during the Straits discussions, the visible deterioration of relations was accompanied by a series of other issues from trade problems to spying conspiracies.

The story of these incidents has been well told.[67] The first major issue was connected to some earlier incidents which related to the arrest and imprisonment of two British citizens, of whom one was executed, for their allegedly 'counter-revolutionary' activities in Russia. They were Mr C. F. Davison, an engineer, and Mrs Stan Harding, a journalist. Davison had been arrested in September 1919 in connection with spying activity in Russia under the directives of Sir Paul Dukes, chief of the British intelligence service in Russia in 1918–19. Davison had been shot four months later.[68] Harding had been accused of espionage and detained for five months in 1920.[69] Both cases had been raised by the British government several times. Britain demanded a satisfactory explanation and full compensation from Moscow.[70]

The second problem was the prosecution of the Russian clergy. Religious prosecution had long been a source of protest in the West. Early in 1922 the Soviet government issued a decree that the sale of church treasures would be undertaken in order to provide for famine relief, which had initially been proposed by a group of churchmen. In some places, however, local clergy opposed the removal of the treasures. The Soviet press started a campaign against the Orthodox Church and the Synod was accused of being the 'general staff of the counter-revolution'.[71] A number of protests were issued in Britain when on 9 May 1922 the patriarch of Moscow, Tikhon, was placed under house arrest in the Donskoi monastery.[72] When on 26 March 1923 a Roman Catholic priest, Budkiewizc, was condemned to death in Moscow for high treason the event became a central issue in a big public campaign in Britain.[73]

In the meantime, Tikhon was arrested but the trial did not take place in 1922. The decision to put him on trial was delayed several times. Chicherin asked Stalin to prevent the execution of Tikhon. He mentioned that the execution of Budkiewizc had harmed the Soviet Union's international position and because of Tikhon's eminent position his execution would cause an even greater furore. He also mentioned that France was considering bringing the case of Budkiewizc before the international negotiations in Lausanne. Tikhon was released after giving a statement to the Soviet press, in which he denounced himself for his 'hostility' to Soviet power and repented his behaviour.

Another disputed area was the question of fishing rights and territorial waters which again was related to a series of events going back to pre-war days. Successive Russian governments had claimed a 12-mile territorial water limit off the northern coasts of Russia. This the successive British governments had refused to recognise. The British had disputed the legitimacy of a 12-mile limit and maintained that a territorial limit greater than three miles was not acceptable.[74] The Soviet government, like its Tsarist predecessors, arrested any foreign trawlers, including British, caught fishing within the 12-mile limit. The dispute regarding the territorial waters was further complicated in March 1922 when a protest was made over the arrest of the crews of two British-owned vessels, the 'Magneta' and the 'St Hubert' off the northern coast of Russia. All the British fishermen arrested by the Soviet authorities were released in early 1923.[75] On 31 March 1923 however, another British steam trawler, 'James Johnson', was arrested off the Murman coast.[76]

In April 1923 Vickers Ltd. asked for permission from the Foreign Office and the Admiralty to sign a contract with the Soviets to sell 500 submarine mines to the Bolshevik government. Vickers indicated that the contract would relieve unemployment in many parts of Britain. Curzon was infuriated by this possibility and by the idea of selling British arms that could kill British soldiers. As a result, on 25 April the Foreign Office made it clear to the Admiralty that 'for the present on political grounds the proposed transaction is considered as highly undesirable'.[77]

Two weeks later on 8 May 1923 (two days before Vorovsky was murdered), a British note had been handed to the Soviet government.[78] The British note, a long memorandum of 26 paragraphs, listed all the aggravations.[79] It included claims arising out of the arrest and persecution of the British citizens, complaints about the detention of British fishing vessels and the trials of the priests. The British note also asked the Soviet government to stop all anti-British propaganda in Asia.[80] The note declared that unless the Soviet government undertook within 10 days to comply fully and unconditionally with the requests which the British note contained, Britain would consider herself free from the obligations of the Anglo–Soviet Trade Agreement.[81] On the day on which Vorovsky was assassinated Roland McNeill, under-secretary for foreign affairs, informed the House of Commons that H.M.S. *Harebell* had been sent to the Archangel district with instructions to prevent interference with British vessels outside three-mile limit, using force if necessary.[82]

The British note, in view of its tone, demands and the 10-day limit gives the impression that the British government was determined to follow a hard-line policy even though it might mean a break up. The British note was criticised as being a provocation, and together with Curzon's attitude towards the Soviet delegation at Lausanne, it was claimed that the British government was pursuing a war policy.[83] When on 10 May, just two days after the delivery of the British note, Vorovsky was shot dead at Lausanne, this conviction was strengthened.

The Bolsheviks considered the 'Curzon note' nothing but a provocation, but they also knew that they could not afford a rupture.[84] To this end Moscow tried to be as conciliatory as possible. The Soviet government promptly replied to the British note on 11 May. On the same day Krasin left Moscow for London to meet British officials. The Soviet reply was very conciliatory in tone, declaring its readiness to discuss the matter of arrested trawlers and the three-mile limit in an international conference. With reference to the charges of anti-British propaganda it was maintained that the British note was based on a combination of fictional, false, altered and arbitrary information. The Soviet reply concluded that the Soviet government was ready to discuss all these points at a conference and determined to keep its relations with Britain on a friendly basis.[85]

In London Krasin had a meeting at the Foreign Office on 17 May where both the British note and the Soviet reply were fully discussed. On 23 May 1923 Krasin presented to Curzon another note, which was in fact a reiteration of the official Soviet reply and again very conciliatory in tone.[86] This was followed by a second British note on 29 May. As it was clear in this second British note, delivered after the negotiations at the Foreign Office, the only remaining dispute was the question of anti-British propaganda in the East. The note specifically demanded that the Soviet representatives in Tehran and Kabul 'will, within a reasonable space of time, be transferred to some other areas where their duties will not bring them into contact with British interests.'[87] Moscow replied to this British note on 4 June 1923.[88] On the question of the recall of Soviet representatives at Kabul and Tehran the Soviet reply stated: 'The Russian government invites the British government to admit that such an exceptional measure, even in the moderated formula of the last memorandum, could be decided upon only as the result of a joint examination of one-sided or mutual accusations.'[89]

The British Foreign Office sent a reply to Krasin on 13 June 1923. It was evident from the tone of the response that both sides were more relaxed although the charges against the Soviet representatives in Kabul and Tehran were still brought forward. 'The Soviet government having thus complied the essential conditions of the demands put forward by His Majesty's Government, this correspondence may now be brought to a conclusion.'[90] Moscow replied on 18 June, welcomed the last British note and agreed that the correspondence may be considered at an end.[91]

Final Agreement

As to the Turkish settlement, the conference at Lausanne continued to discuss terms for a long-term settlement in the absence of the Soviet delegation. The rest of the conference turned around the questions of the Thracian boundary, the Aegean islands, the Turkish–Iraq boundary in the Mosul region, the problem of minorities and the issue of capitulations involving judicial, financial and economic elements. The district of Mosul, with its rich deposits of oil and mainly Kurdish inhabitants, proved the most difficult question as here there existed a fundamental conflict between the vested interests of Turkey and Britain. Both delegations considered a favourable settlement of this matter to be of the greatest importance.

The Mosul question raised political and economic questions of the highest order. The crux of the Mosul impasse into which Turkey and Britain drifted was conditioned by the question of oil, as well as by territorial claims. No solution could be reached to this dispute during the conference. The question of Mosul was thus postponed and left to direct negotiations between Britain and Turkey. A settlement was reached only after three years in June 1926 with the signing of a treaty recognising the mandate of Iraq.[92] In this way Turkey had to accept the decision of the Council of the League of Nations with some minor modifications and had to withdraw its claim on the region for good.[93]

Although less intractable than the question of Mosul another major issue was the region of Thrace, for it dealt with the populations, and the future borders of eastern and western Thrace. By the Lausanne Treaty Turkey received full sovereignty over eastern Thrace but had to accept demilitarised zones protecting the Greek and Bulgarian frontiers. Turkey was allowed troops in eastern Thrace and was awarded the town of Karaagaç. The conference also agreed that a mutual exchange of Greek and Turkish populations in the reconstituted areas would take place. Over a million Greeks were forced to leave in return for 350,000 Turks expelled from Greece.[94]

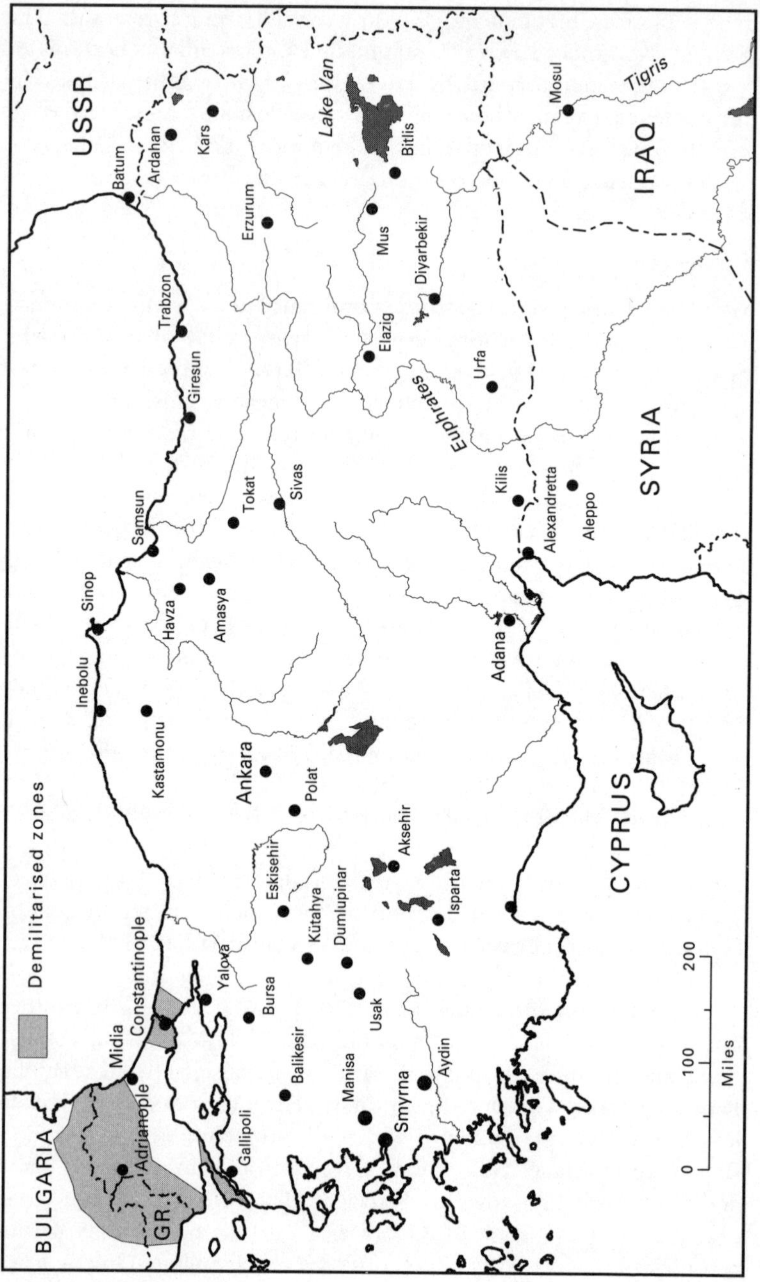

Map 9 Turkey after Lausanne

Considerable debate developed over the ownership of various Aegean and Dodacenese islands. Finally it was decided by the conference that Tenedos, Imbros, and the Rabbit Islands would go to Turkey because of their proximity to the Straits. The Dodacenese islands would go to Italy, and the rest of the Aegean islands would remain in Greek territory.[95]

Emotionally, the problem of the capitulations proved more difficult to resolve. These were extra-territorial and supra-national privileges granted to foreigners by the Ottoman Empire for centuries in the spheres of justice, commerce and finance. As the Ottoman Empire weakened and other nations were granted more and more capitulations, those privileges covered almost every aspect of life engaged in by foreigners residing in the Ottoman Empire.[96] The Turks viewed them as the most serious infringement of their sovereignty. The greatest difficulty were the judicial capitulations which placed the foreigners outside of the Turkish civil law and made them subject to the laws of their respective countries. The Allies consented in principle to the abolition of capitulations but insisted that some special arrangements must be made for the time being until the new Turkish state created an acceptable judicial system.

When the final treaty was signed the capitulations were abolished, and foreign residents, businesses, economic concessions and schools were subjected to Turkish law. The judicial issue was settled by providing merely for notification of arrests of foreigners immediately following the arrest. Turkey agreed to keep some foreign judicial advisors for the next five years as a transitory measure.[97]

Despite the end of capitulations the section on economic clauses included certain five-year restrictions on customs duty, contracts and financial concessions, all subject to review in disputed cases by a mixed arbitration tribunal. Regarding the question of the Ottoman debt an agreement was eventually reached in July 1923 fixing it at 141 million Turkish liras, of which Turkey was to be responsible for 84 million and the successor states in the Middle East for the remainder.[98] Finally on reparations, no reparation was paid except the Turkish gold already in Allied hands.[99]

Conclusion

Despite the fact that the conduct of the British delegation at Lausanne was hailed as a triumph,[1] the Lausanne treaty was not an overall victory for the British. On the contrary, what occurred at Lausanne conclusively buried the immediate post-war dreams of the British Empire to establish a regional structure on the basis of its First World War victory in the Near East. Although at the conference the British delegates achieved most of their aims – as opposed to the Soviet and some of the Turkish demands – these had already been very much modified and were almost completely different from the original British aims in the immediate aftermath of victory at the war fronts in 1918. In Toynbee's words, it was merely 'face-saving on the part of the nominally victorious vanquished'.[2]

At the end of the war it had seemed to many in Britain that a regional system controlled by the British Empire could be successfully established. This regional system was to consist of an enlarged Greek state extending to western Anatolia, an autonomous Kurdish state in eastern Anatolia, and the independent states of Armenia, Georgia and Azerbaijan in the Caucasus. The new Russian government in Moscow had not been considered as an important factor at that stage. Neither had there been anticipated any serious resistance from Turkey to this grand design.

By 1923, however, the scenario projected five years earlier had failed to materialise in the region. The Greek campaign had failed completely and an independent Turkish state was founded in Anatolia and eastern Thrace. The independent states of Transcaucasia had been captured by the Bolsheviks. The deviation from the original intention of the British Empire was confirmed in a British Foreign Office Memorandum on 15 November 1922:

> Since the signature of the Treaty of Sèvres [August 1920] the circumstances under which that Treaty was drafted have been radically altered. Turkey has recovered; Russia has again become a serious menace, and the

Allied Powers have failed, on several important occasions, to maintain a united front.

The memorandum concluded that it would 'not only be almost impossible ... but it would prove physically difficult to maintain the former British position' due to changed circumstances.[3] The Lausanne agreement represented the abolishment of 'the extreme "British" point of view' which 'was represented by the Treaty of Sèvres'.[4]

The Lausanne settlement may well be regarded as one of the major turning-points in the history of Anglo–Russian relations, for it facilitated an effective finale to a long-standing problem around which the relations between two sides had revolved for most of the nineteenth and early twentieth centuries. This was the Eastern Question which had its origins in the disintegration of the Ottoman Empire.

From at least the 1830s onwards, the international powers that were mainly concerned with, and for most of the time directly involved in, the affairs of Turkey were Britain and Russia. The region was of first importance for both imperial powers. They were engaged in rivalry for prestige, influence and material interests. Interrupted by the emergence of German power which was aspiring to acquire regional influence in the Near East, the familiar pattern of Anglo–Russian rivalry had resurfaced again in the aftermath of the war. First the events of the Caucasus and then the Greco–Turkish war provided a new framework for the old rivalry. This severe conflict between Soviet Russia and Great Britain constituted a decisive factor for the future of the region.

British involvement in the Caucasus came in late 1917 to fill the apparent vacuum caused by the disintegration of the Russian armies following the Bolshevik Revolution. It was considered a preventive action aimed to protect the Caucasus against the German and Turkish armies. However, the British involvement in the region, for the most part, remained ineffective taking the form of material and moral support provided to anti-Bolshevik Russian forces. The policy can best be described as low-cost control for the purposes of defence, trade and imperial grandeur. The Caucasus was never considered as indispensable to the British as it was to the Bolsheviks. When it became obvious that more resources were required to maintain an effective control the British preferred to withdraw leaving the region to the Bolsheviks.

The main objective of the Soviet government was to secure the Soviet power in the borderlands of ex-Tsarist Russia. The Caucasus was particularly important in this effort as it was considered the most vulnerable borderland of the Soviet state. The vital ingredients of the Soviet Caucasian policy were all to be found in the British unwillingness to commit themselves to an expensive involvement in the region. Soviet policy makers concluded that the Western resistance to the Soviet southern move in the region would be weak or non-existent. When British forces finally withdrew in the summer of 1920 the Bolsheviks were ready to step into their place.

With the Caucasus firmly placed in the Soviet sphere of influence, the focus of the rivalry between two powers shifted to the south-west. During the next three years the rivalry between Britain and Soviet Russia found a new outlet in the context of the ongoing Greco–Turkish war, with Britain standing behind the former while the Soviet state providing material help and moral assistance to the latter.

British policy in Turkey, like that in the Caucasus, had initially been based more on perception of opportunity than on perception of threat. After the consolidation of the Caucasus within the Bolsheviks' sphere and the emergence of the Turkish national movement as a serious challenger, the policy responded to the new circumstances and became defensive, aimed at protecting the imperial life-line against a united Turco–Bolshevik attack. In this way defence, rather than further expansion of the empire became uppermost in the minds of those officials responsible for administering British policy.

Soviet policy in Turkey, in broader terms, seems to have been subordinated to the primarily negative objective of weakening the influence of British power adjacent to its southern borders. Adopting the role of a balance in the entangling Anatolian struggle between the Turkish nationalists and the Greek army, Moscow was determined not to allow the implementation of British designs.

Despite differences in style, Soviet policies in Turkey were just as pragmatic as British policies. Soviet attitudes during the period under examination present an unending balancing act between militancy and moderation. On the one hand, there was the cry of the local communists, which from time to time found support in the agitation of the Comintern. On the other, there was growing evidence that Moscow was not after a revolutionary change in Turkey. The main reason for this was related to a simple practical calculation: the communist movement in Turkey, though sufficiently powerful to create confusion in the country, was not strong enough to take over the reins of power. In view of this, Moscow could hardly rely exclusively on the local communists to serve as an effective counter-balance to British influence in the region. The Soviet leadership, while by no means rejecting the concept of the inevitable triumph of world communism at some future point in time, came to the conclusion that the interests of the Soviet state could best be served by friendly relations with the 'bourgeois nationalist' Kemalist regime of Turkey.

Optimism in local communist circles regarding revolutionary prospects in Turkey was not only unwarranted, in retrospect it was also self-defeating. Their judgement was misguided about the general attitude towards the communist movement in Anatolia when the British occupation led to a widespread resentment against the Western Powers. At no time during the period under review did the general conflagration foreseen in the region become imminent in reality. The tragic end of Mustafa Subhi and the other leading Turkish communists was a striking example.

The Soviet state, though seeking a viable counter-balance to the British presence in Turkey, was not after a deep involvement in the intractable problems of Anatolian conflict that were of no direct concern to vital Soviet interests. In Turkey, by contrast to the Caucasus, Soviet security was not directly at stake. As a result Moscow used its support to Ankara in a subtle way, showing an ability to minimise the damage to its interests abroad while applying its capabilities incisively. The military support given to Turkish nationalists was kept at a level easy to deny, and the British political scene was watched carefully. The Bolsheviks showed restraint and avoided agitating the Turks any further than necessary. The overall Soviet strategy was dictated by the desire not to jeopardise their international position.

Both the British and Soviet involvement in Turkey after 1920 was on the whole based upon foreign policy aims which were essentially defensive. In this complicated and rather explosive regional crisis both sides tried to play safe to avoid further commitments which would surpass the capabilities of both. Britain and the Soviet state did not want a direct military confrontation in the region, and indeed there was no necessity for one. In the end a fortuitous circumstance was created in the region that neither of these powers was either strong enough or determined enough to commit itself to fight for ultimate control.

The role of the Turkish resistance was far from that of a passive observer of Great Power intrusion into Turkish affairs. At first the conditions were against any chance of possible resistance to the Allied policies. In Toynbee's words:

> Turkey's provinces were gone; her allies were crushed; and except for her champions among the Indian Muslims, she was friendless even in the camp of Islam. Constantinople was held by the victors, Turkey was encircled by enemies. Like wolves around the camp fire the Powers were prowling at the threshold with hungry eyes, for Turkey by nature is rich, and imperialism is greedy.[5]

Within this hopeless situation the leaders of the national movement utilised every possible opportunity presented by the post-war circumstances. They used their ethnic and religious prestige among the Muslim population of the Caucasus to increase their credibility in the eyes of the Bolsheviks. Turkish nationalists made the most of these early contacts with the Bolsheviks in the Caucasus as a testing ground, exploring alternative courses of action and securing material and moral assistance from the young Soviet state. In this way they broke their isolation and acquired a material basis on which to organise military resistance in western Asia Minor.

When the British forces withdrew from the Caucasus it was only a matter of time before the two sides that bordered the Caucasus would re-establish their hegemony. As it turned out, the bulk of historic Armenia fell to the Turkish nationalists, while eastern Armenia, Azerbaijan and Georgia came under Soviet control. This could not but give an immense psychological satisfaction and moral encouragement to the Turkish nationalists demonstrating that the supposedly formidable British Empire was not so formidable at all. With the Caucasian borders established firmly between the Turks and the Bolsheviks, the Turkish nationalist forces concentrated on western Anatolia to wage war against the Greek occupation. They fought well, particularly in 1922; the Turkish army, after the initial confusion and low morale, regained its self-confidence to a large degree.

On the diplomatic front, the Turkish nationalists exploited the divergence of policy within the Allied camp and the antagonism between Soviet Russia and Britain. They soon found out that their goal could be achieved only if the Bolsheviks and the British were balanced against each other. Turkey could remain the buffer between the two rival powers and thus preserve its national existence. In order to implement this strategy they accepted military help and diplomatic assistance from Moscow without committing themselves to the cause of the Bolsheviks. They were determined to fight back against the plans of the British and their allies in Turkey without abandoning the diplomatic contacts with the Allies. If such a painstaking path in foreign policy had not been followed the military victory could never have gained the same effective results. In the end, the independence of Turkey was safeguarded 'as securely as possible between Russia on the North and Great Britain on the South'[6].

With the Lausanne settlement the lands of Turkey ceased to be a major cause for the ambitions and rivalries of the Great Powers. The Eastern Question was no more. International tension in the region was reduced to a minimum. For a considerable length of time peace reigned in this region.

Notes

Introduction

1 The terms 'Middle East' and 'Near East' have been used in this text as general short-hands, the 'Near East' comprising Greece, Bulgaria, Turkey, and the Levant (the eastern coast of the Mediterranean and its hinterland), and the 'Middle East' comprising Arabian Peninsula, Mesopotamia, the Persian Gulf and Iran.

2 Discussions on the captured German colonies, Imperial War Cabinet Minutes, 20 December 1918; London, PRO; Cab. 23/42.

3 With the exception of Mosul *vilayet* that was controlled by the Turkish armies under the command of Ali Ihsan Pasha. Only after the armistice, the British forces in Mesopotamia occupied Mosul and the surrounding area.

4 The best general survey of the 'Eastern Question' is Anderson's *Eastern Question 1774–1923. A Study in International Relations.*

5 For the history of the Straits question see Jelavich, *The Ottoman Empire, the Great Powers and the Straits Question, 1870–1887*; Puryear, *England, Russia, and The Straits Question* and Howard, *The Problem of the Turkish Straits.* For a useful short Russian summary of the issue see Miller, *Turtsiia i Problema Prolivov.*

6 Tukin, *Bogazlar Meselesi*, pp. 65–107. This is the most comprehensive work in Turkish on the Straits question.

7 'Constantinople' has been chosen over 'Istanbul' because it was the name frequently used in the Ottoman documents, and most often used in European languages as well as in Russian, during the period under consideration.

8 Longrigg, *Oil in the Middle East*, p. 27; Earle, *Turkey, The Great Powers and the Baghdad Railway: A Study in Imperialism*, pp. 70–84.

9 The Anglo–Russian Convention is conveniently examined in R. P. Churchill's *Anglo–Russian Convention of 1907.*

10 Until February 1918 (according to the Gregorian calendar which was/is in use in Western Europe) Russia used the Julian calendar. All the dates in the Julian calendar are 12 days before in the 19th century and 13 days before in the 20th those of the Gregorian calendar. All dates in this text are given in accordance with the Gregorian calendar.

Chapter 1

1 Lenin, *Collected Works*, XXVI, p. 62.

2 Lenin, 14–15 October 1917, *Collected Works*, XXVI, p. 137.

3 Taylor, *The Struggle For Mastery In Europe*, p. 559.

4 The Decree on Peace was passed by the Second All-Russian Congress of Soviets of Workers' and Soldiers' Deputies which met in Petrograd on 7 and 8 November, 1917. (Lenin, *Collected Works*, XXVI, pp. 243–63.)

5 Lenin did not call for the abolishment of all the secret treaties. Although he declared the abolishment of secret diplomacy he clarified himself by saying that 'we reject all clauses on plunder and violence, but we shall welcome all clauses containing provisions for good-neighbourly relations and all economic agreements, we can not reject these.'(*Collected Works*, XXVI, p. 255.)

6 *DVP*, I, pp. 11–14. *Pravda*, 28 November 1917, p. 1. An English translation in Lenin, *Collected Works*, XXVI, pp. 249–53.

7 Lloyd George, *War Memoirs*, II, pp. 1544–52.

8 Clayton, *The British Empire as a Superpower, 1919–39*, p. 46. Churchill writes that there were 80 divisions of German and Austrian troops upon the Russian front in October 1917. (Churchill, *The World Crisis, Part V: The Eastern Front*, p. 353.)

9 The essential work on the beginning of the foreign intervention in Russia is Ullman's *Intervention and the War* (chapter II, pp. 40–57.) which presents a thorough investigation of various British and other Allied initiatives.

10 The War Cabinet, a compact group of from five to seven members, had been formed by Lloyd George in December 1916 to respond to the numerous wartime problems. The Imperial War Cabinet composed of the War Cabinet and representatives, usually the premiers, of the British dominions and of India.

11 George, *War Memoirs*, II, pp. 1446, 1540.

12 'Voisko Cossack Circle', a congress which had been elected, over two hundred years before, by the Don Cossacks as a nation. 'Ataman' was traditionally elected at a general assembly or circle and in time of peace had merely the function of presiding over the assembly, whereas in time of war he had the absolute authority. General Kaledin was elected as Don Ataman in June 1917. (Starikov and Medvedev, *Philip Mironov and the Russian Civil War*, pp. 27–9.)

13 Writing in September 1917, Lenin accused Kaledin and Kornilov for 'moving toward civil war'. *(Collected Works*, XXVI, pp. 33–4.)

14 Starikov and Medvedev, *Philip Mironov and the Russian Civil War*, pp. 36–7.

15 Summers, American Consul General at Moscow to the Secretary of State, 8 November 1917, *Foreign Relations, 1918, Russia*, II, p. 580.

16 Alikberov, *Pobeda sotsialisticheskoi revoliutsii v Dagestane*, pp. 51–2. An English translation of the Treaty of the Union in *Foreign Relations, 1918, Russia*,

II, pp. 616–18.

17 Lloyd George, *War Memoirs*, II, pp. 1544,1546 and 1551; From Smith, American Consul at Tiflis to Lansing, Secretary of State, 18 November 1917, *Foreign Relations, 1918, Russia*, II, p. 581.

18 ibid.

19 From House(Paris), the Special Representative, to the Secretary of State, 2 December 1917, *Foreign Relations, 1918, Russia*, II, pp. 583–4.

20 'Notes on the Present Russian Situation' by A. J. Balfour, 9 December 1917, in Lloyd George, *War Memoirs*, II, p. 1547.

21 ibid., p. 1548.

22 Kaledin's regime, however, did not survive long. At first Kaledin's initiative attracted further support from some other well-known Tsarist generals like Kornilov, Alexeiev, Denikin and Lukomsky. In early January the latter four started to organise a 'Volunteer Army' and there were attempts to set up a united movement together with Kaledin. In the second half of January, however, most of the Cossack regiments which had recognised the authority of Kaledin left him. As a result of the insurrectory Cossacks in the Don region the Volunteer Army decided to leave the Don region and go to the Kuban. Considering the situation hopeless, Kaledin shot himself on 29 January. (Starikov and Medvedev, *Philip Mironov and the Russian Civil War*, p. 52.)

23 From Page(London) to the Secretary of State, 21 December 1917, *Foreign Relations, 1918, Russia*, II, pp. 595–6.

24 When the Second All-Russian Congress of Soviets voted itself into power on 7 November 1917 an All-Bolshevik Council of People's Commissars took over the government. This organ was to be responsible to the All-Russian Congress of Soviets and to its Central Executive Committee.

25 The message was received at Field Headquarters in the early morning on the following day. Not having received a reply from Dukhonin by the evening of the same day, Lenin, Stalin and Krylenko started a two and a half hour conversation with Dukhonin over the direct line. When, during the conversation, Dukhonin refused to enter into formal negotiations for an armistice, he was informed that he was dismissed from his post, and Krylenko was appointed the new commander-in-chief. (Trotsky, *Stalin*, pp. 245–6 and Lenin, *Collected Works*, XXVI, pp. 312–3 and 316–7.)

26 Order of the Central Executive Committee to General Dukhonin, Russian commander-in-chief, 21 November 1917, *DVP*, I, pp. 15–16. English translation of it in Degras, *Soviet Documents on Foreign Policy*, I, 3–4.

27 Address from the Soviet Commissariat of Foreign Affairs, 8(21) November 1917, *DVP*, I, pp. 16–17; English translation in Degras, *Soviet Documents on Foreign Policy*, I, 4.

28 Extracts from a statement by Trotsky on the International Situation at the Central Executive Committee, 21 November 1917, Degras, *Soviet Documents on Foreign Policy*, I, p. 6.

29 Russia had entered into the engagement in September 1914 by the Treaty of London not to make a separate peace. (Lloyd George, *War Memoirs*,

II, p. 1543.) In April 1915, Great Britain, France, Italy and Russia signed a declaration in London, accepting 'not to conclude a separate peace during the course of the European War.'(*British and Foreign State Papers*, 1919, CXII, pp. 977–8.) This treaty, together with the other war-time agreements concluded by the Tsarist and later by the Provisional government of Russia with the governments of Britain, France, Germany, Japan and other powers, were published in *Pravda* and *Izvestiia* in November and December 1917. (Lenin, *Collected Works*, XXVI, pp. 552–3.)

30 Statement by Trotsky on the publication of the secret treaties, 22 November 1917, Degras, *Soviet Documents on Foreign Policy*, I, p. 8.

31 Telegram from the US Ambassador in Russia (Francis) to the Secretary of State, 27 November 1917, *Foreign Relations, 1918, Russia*, I, p. 250; Lloyd George, *War Memoirs*, II, p. 1542.

32 Lloyd George, *War Memoirs*, II, pp. 1540–3; Knox, *Russian Army*, II, p. 727; Buchanan, *Mission*, II, p. 225.

33 Lloyd George, *War Memoirs*, II, p. 1542; Telegram from Colonel House to the President, Paris, 2 December 1917, Seymour, *The Intimate Papers of Colonel House*, III, pp. 290–1.

34 'Supplies for Russia' (War Office Accounts forms), 22 and 28 November and 16 December; London, PRO; FO 371/3347.

35 In Lloyd George, *War Memoirs*, II, pp. 1550–1. Ullman misquotes this sentence replacing 'quietly' by 'quickly'. *(Intervention and War*, p. 54.)

36 The text of this convention is printed in French in *British Documents*, III, pp. 369–70, and an English translation of it in *Foreign Relations, 1918, Russia*, I, pp. 330–1.

37 Before the First World War, the Ukraine had been the 'granary' of Russia. Nine-tenths of all the cereals exported by Russia came from the Ukraine, which sent out also 230,000 head of cattle per year. (Cited in Conte, *Christian Rakovski (1873–1941)*, p. 155.)

38 ibid.; Lloyd George, *War Memoirs*, II, pp. 1550–1.

39 The concept's most influential intellectual exponent was Ziya Gökalp, critic, publicist, and later professor of Sociology at Istanbul University. In his poem 'Turan', published in 1911, Gökalp summarised his Pan-Turanian vision as

> The country of the Turks is not Turkey,
> nor yet Turkistan,
> Their country is a vast and eternal region: Turan!'
> (Heyd, *Foundations of Turkish Nationalism*, p. 126.)

40 Lloyd George, *War Memoirs*, II, 1551.

41 Aydemir, *Makedonya'dan Orta Asya'ya Enver Pasa*, III, pp. 392–4. This is an important and authoritative first-hand account of S. S. Aydemir who was in the Caucasus and Russia both during and after the war.

42 Lord Alfred Milner pointed out the danger of Turkish expansion at the beginning of 1918. (Given by Stanwood, *War, Revolution and British Imperialism in Central Asia*, p. 63.)

43 The Convention in *British Documents*, III, pp. 369–70.

44 Ian Malcolm, 'Diary of Russian Affairs From the Revolution to Peace with Germany', *British Documents on Foreign Affairs*, II, series A, pp. 8–21.

45 Telegrams from Page(London) to the Secretary of State, 18, 19 and 21 December 1917, *Foreign Relations, 1918, Russia*, II, pp. 591–6.

46 Lord Milner at the beginning of 1918, in Stanwood, *War, Revolution and British Imperialism*, p. 63.

47 In early days of 1918 it was also decided to encourage and assist the elements in Siberia to resist a German invasion of the region. This was followed the arrival of Japanese warships on 12 January, and a British cruiser two days later in Vladivostok. (From the Director of Military Intelligence to Under Secretary of State for Foreign Affairs, 17 March 1918; London, PRO; FO 371/3290.) A detailed account of the Allied intervention in Siberia in Ullman, *War and the Intervention*, pp. 90–103.

48 *Istoriia Vneshnei Politiki SSSR*, I, p. 31.

49 Bondarevskaya, 'Velikii Oktiabr' i Krizis Kolonial'noi Politiki Velikobritanii Na Vostoka', *Velikii Oktiabr' i Istoricheskii Progress*, pp. 269–70.; Trukhanovsky, *Winston Churchill*, p. 147.

50 'Pasha' was a title placed after the name borne by civilian and military officials of high rank. In the military establishment it was equivalent to the rank of general.

51 Talat Pasha's statement, *Tanin*, 5 April 1917, given by U. Kocabasoglu in his unpublished manuscript, *Rus Devrimi ve Osmanli Basini*, p. 33.

52 *Tasvir-i Efkar*, 6 June 1917, ibid., pp. 56–7.

53 *Tanin*, *Sabah*, *Tasvir-i Efkar*, 12 November 1917, ibid., p. 91.

54 Ludendorf, *My War Memoirs 1914–1918*, II, p. 620.

55 The *vilayets* were large administrative divisions. In 1864 the empire had been divided between 27 of them. The *vilayet* was divided into *sanjack*; the *sanjack* into *kazas*; the *kaza* into *nahiyes* and villages. (Lewis, *The Emergence of Modern Turkey*, p. 388.)

56 Yerasimos, *Türk–Sovyet Iliskileri*, pp. 17–19; Duran, 'I. Dünya Savasi Sonunda Türk Diplomasisinin Ilk Basarisi: Brest–Litovsk Hazirliklari', *Belgelerle Türk Tarihi Dergisi*, April–June 1973, p. 26; Gratz and Schuller, *The Economic Policy of Austria–Hungary during the War*, p. 114.

57 Ahmad, *The Making of Modern Turkey*, p. 46. This is one of the recent books on modern Turkish history which raises interesting and difficult questions of interpretation of the period.

58 Enver (1881–1922) was one of the leading members of the CUP, the ruling party during the war, and son-in-law to the 'shadow of God on earth' (the caliph in Constantinople). He became the most powerful person in the state in 1913 when he became minister of war. Led the reorganisation of the Ottoman Army and the Ottoman war effort in World War I. (Aydemir, *Makedonya'dan Orta Asyaya kadar Enver Pasa*)

59 Cebesoy, *Milli Mücadele Hatiralari*, p. 42; Karabekir, *Istiklal Harbimizde Enver Pasa ve Ittihak Terakki Erkani*, p. 4.

60 Telegram from the Minister in Sweden (Morris) to the Secretary of

State, File no. 763. 72119/1430, *Foreign Relations, 1918, Russia*, p. 434.

61 According to this treaty, Russia lost 34% of its population, 32% of its agricultural land, 54% of its industrial undertakings, and 89% of its coal mines. (*Foreign Relations, 1918, Russia*, I, 490.)

62 'Peace Treaty between Russia, Germany, Austro–Hungary, Bulgaria and Turkey', 3 March 1918, *DVP*, I, pp. 119–24.

63 The Turkish–Russian Supplementary Treaty at Brest–Litovsk was signed on 3 March 1918. (From Chicherin to Galip Kemali Bey, Ottoman representative in Moscow, 23 May 1918, Moscow, AVP, Fond: Reference about Turkey, Op.: 1, Por.: 2, Pap.: 1.) G. V. Chicherin (1872–1936). Member of Tsarist Ministry of Foreign Affairs, resigned in 1904, emigrated to Berlin where he joined Mensheviks. In Britain during First World War. Arrested in late 1917, exchanged for British ambassador, Sir George Buchanan. Returned to Russia in January 1918, replaced Trotsky as commissar for foreign affairs. (Chicherin's authorised biography in *Entsiklopedicheskii Slovar Russkago Bibliograficheskago Instituta Granat*, XLI, Part III, p. 215; For a good English biography of Chicherin see O'Connor's *Diplomacy and Revolution*)

64 General accounts of the Turkish gains at the Brest treaty include Wheeler-Bennett, *Brest–Litovsk*, p. 272 and Miller, *Ocherki Noveyshey Istorii Turtsii*, pp. 66–7.

65 Daghestan is a Turkic term meaning mountain country.

66 Kurat, *Türkiye ve Rusya*, pp. 671–2; Gugov, *Sovmestnaia bor'ba narodov Tereka za Sovetskuiu vlast'*, p. 192; Korenev, *Revoliutsiia na Tereke, 1917–1918 gody*, pp. 53–4.

67 From the statement issued by the Georgian Information Bureau, London, 26 June 1918, *Foreign Relations, 1918, Russia*, II, pp. 634–7.

68 *Bor'ba za pobedu sovetskoi vlasti v Gruzii. Dokumenty i materialy(1917–1921)*, pp. 7–8; Kazemzadeh, *Struggle for Transcaucasia*, p. 55; Hovannisian, *Armenia on the Road to Independence, 1918*, p. 107.

69 The closest to a democratic election ever held in Russia, elections to the Constituent Assembly took place in late November 1917. Of the 707 elected members of the assembly the SRs had a comfortable majority of 410. The Bolsheviks received much less than half of the popular vote and could secure less than a quarter of the seats, i.e. 175. (Carr, *The Bolshevik Revolution*, I, p. 120.)

70 The list of candidates to the All-Russian Constituent Assembly was recognised as the lists of candidates to the *Seim*. The *Seim* consisted of representatives of some ten political parties: Mensheviks, Kadets, SR's, Dashnaks, Social-Federalists, National-Democrats, Musavatists, Ittihadists, Moslem Socialists, and Himmetists. The Mensheviks and the Musavat were the largest factions of the *Seim*, holding thirty seats each. The Dashnaks held 27 seats. Other parties had an insignificant number of deputies. (*Dokumenty i materialy po vneshnei politike Zakavkaz'ia i Gruzii*, pp. 27–8.)

71 Yerasimos, *Türk–Sovyet Iliskileri*, p. 19.

72 Semenov, Iulii Federovich. 'Zakavkazskaia republika', Vorozhdenie. Paris, Jan. 1949. (A chapter from unpublished memoirs of the only Kadet

member of the Transcaucasian Seim.), p. 123, given in *US Foreign Relations, 1918, Russia*, II, pp. 619–20.

73 Gadzhiev, *Iz Istorii Obrazovaniia i Paderiia Iugo-zapadnoi Kavkazskoi (Karskoi) Demokraticheskoi Respubliki*, pp. 6–7. This is an important booklet as the writer has made an extensive use of the state archives of the Azerbaijani Republic.

74 Kurat, *Türkiye ve Rusya*, pp. 467–73.

75 Yerasimos, *Türk–Sovyet Iliskileri*, pp. 21–2; Swietochowski, *Müslüman Cemaatten Ulusal Kimlige Rus Azerbaycani, 1905–1920*, p. 167.

76 Stepan Shaumian (Tiflis 1878-desert near Ashkabad 1918). Formed the Union of Armenian Social Democrats in 1902. Co-founder of *Proletariat* (Tiflis) in the same year. Directed oil-workers' strike May–June 1914. Arrested in 1916 and exiled to Saratov. (Mints, *God 1918*, p. 451–77.)

77 Suny, *The Baku Commune 1917–1918*, pp. 72–5.

78 According to the 1916 census returns, there were 95,000 Muslim Azeris, 90,000 Russians, and 63,000 Armenians in the city. (Given by Suny, *The Baku Commune 1917–1918*, p. 20.)

79 Lenin, *Collected Works*, XXXV, pp. 76, 342, 345, 576.

80 See Swietochowski, 'National Consciousness and Political Orientations in Azerbaijan, 1905–1920', *Transcaucasia*, ed. R. G. Suny, pp. 217–23.

81 Special cavalry units of Muslims from the Turkestan and the Caucasus were formed and incorporated into the Russian Army early in the war. The largest was the Wild Division. (Tschebotarioff, *Russia, My Native Land*, p. 101.)

82 Shaumian, *Izbrannye proizvedeniia*, II, p. 209. Lenczowski gives a much larger number, 10,000 Muslim deaths. *(The Middle East in World Affairs*, p. 72.) The Muslim newspaper *Azerbaijan* gives the number of Muslim deaths as 16,000. (8 October 1918)

83 Mitrokhin, *Failure of Three Missions*, p. 16; Mikoyan, *Memoirs of Anastas Mikoyan*, I, p. 107.

84 ibid., p. 109. The Caucasian Red Army consisted of 20,000 men, of whom 60 to 70% were Armenians. The chief of staff was the *Dashnak* Z. Avetisov. Most of his officers were also *Dashnaks*. (In Suny, *The Baku Commune 1917–1918*, p. 270.)

85 ibid., p. 110. Bolsheviks were the largest group in the Baku Soviet with 58 delegates, followed by 47 Right SRs, 46 Left SRs, 36 Dashnaks, and 28 Mensheviks. (Suny, *The Baku Commune 1917–1918*, p. 231.)

86 Mitrokhin, *Failure of Three Missions*, p. 17.

87 From Vehip Pasha to Akaki Chkhenkeli, the President of the Transcaucasian Republic, 28 April 1918, given in Yerasimos, *Türk–Sovyet Iliskileri*, p. 22; Mints, *God 1918*, p. 454.

88 *Guberniia, Uezd, volost.* Russian administrative territorial units. The largest of these was the *guberniia*, divided into *uezds*, which in turn were subdivided into *volosts.* This system continued under Soviet power until the introduction of the new system of administrative territorial divisions in 1929–30. (Lenin, *Collected Works*, XXVI, p. 230 f.n.)

89 For Batum Conference see Kurat, *Türkiye ve Rusya*, pp. 474–76; *Dokumenty i materialy po vneshnei politike zakavkaz'ia i gruzii*, pp. 313–16.

90 In Hovannisian, *Armenia on the Road of Independence*, p. 195.

91 German army had employed Georgian volunteers in their campaign in Armenia in 1915 and 1916. In so doing the Germans had come in contact with 'influential' Georgians. (Ludendorff, *My War Memoirs*, II, pp. 620–1.)

92 ibid., p. 621.

93 Major General G. M. W. Macdonough, the director of Military Intelligence, informed the Imperial War Cabinet on 27 June of the arrival of von Kressenstein in the Caucasus; London, PRO; Cab. 23/41. See also Ludendorff, *My War Memoirs*, II, p. 659; and Bunyan, *Intervention, Civil War and Communism in Russia*, p. 51.

94 From the statement issued by the Georgian Information Bureau in London on 26 June 1918, *Foreign Relations, 1918, Russia*, II, pp. 634–7. Agreement between Georgia and Germany in Bunyan, *Intervention, Civil War and Communism in Russia*, p. 53.

95 From the American Consul at Tiflis, Smith, temporarily in Moscow, to the Secretary of State, 16 June 1918, *Foreign Relations, 1918, Russia*, II, pp. 632–3.

96 Kadishev, *Interventsiia i grazhdanskaia voina v Zakavkaz'e*, p. 64; Hovannisian, *Armenia on the Road of Independence*, p. 199.

97 With the Berlin Treaty of 1878 Russia had secured Kars, Ardahan and Batum. (Hurewitz, *Diplomacy in the Near and Middle East*, pp. 189–91.)

98 Yerasimos, *Türk–Sovyet Iliskileri*, p. 25; Kurat, *Turkiye ve Rusya*, p. 664.

99 Yerasimos, *Türk–Sovyet Iliskileri*, pp. 27–8.

100 ibid., pp. 29–30; also see Ziemke, *Die neue Türkei: Politishche Entwicklung*, pp. 54–5, and Hovannissian, *Armenia on the Road of Independence*, p. 174.

101 Kipling and Dunsterville had shared a study at the United Services College, Westward Ho in 1880. (Carrington, *The Life of Rudyard Kipling*, p. 22.)

102 Ranald MacDonell writes in his memoirs that there were a number of British Intelligence Officers in the Caucasus, whose duties 'were to prepare, devise or create a situation which would enable General Dunsterville and his force to take charge.'(MacDonell, *And Nothing Long*, p. 205.)

103 Moberly, *History of the Great War Based on Official Documents, the Campaign in Mesopotamia, 1914–1918*, IV, pp. 104–5.

104 Dunsterville, *The Adventures of Dunsterforce*, pp. 11–67.

105 Major Goldsmith describes his activities with the Caucasus Military Agency in his 86 page report to the director of Military Intelligence. Goldsmith's report reads like a political thriller at times with all the required ingredients of intrigue, conspiracy and frustration. Goldsmith spent eight months in the region until 6 October 1918 when he and the other members of the Caucasus Military Agency were arrested by the Bolsheviks in Vladikavkaz. (From Major G. M. Goldsmith, Acting British Military Agent, Army of Caucasus Front to the Director of Military Intelligence, WO, 1 July 1919; London, PRO; WO 95/4960.)

106 From Sir C. Greene(Tokyo) to Balfour, 12 June 1918; London, PRO; FO 371/3292.

107 Moberly, *History of the Great War*, p. 179; Dane, *British Campaigns in the Nearer East. 1914–1918*, II, p. 215. (Edmund Dane was the Military Correspondent of the *Westminster Gazette*.)

108 Goldsmith's report to the Director of Military Intelligence, 1 July 1919; London, PRO; WO 95/4960.

109 Moberly makes a few general references to the activities of the Caucasus Military Agency. (*History of the Great War*, pp. 103, 104–5.)

110 ibid., pp. 172–3.

111 ibid.

112 ibid., p. 179.

113 ibid.

114 The Eastern Committee was established in March 1918 by the War Cabinet to administer the British policy over a vast area between the eastern Mediterranean and the frontiers of India. Lord Curzon chaired the Eastern Committee and membership included General Smuts of South Africa, Balfour, Cecil (assistant foreign secretary), Lord Hardinge (permanent under–secretary at the FO), Henry Wilson (chief of imperial general staff), Montagu (secretary of state for India) and Major General Macdonough (director of military operations). (Ullman, *Intervention and the War*, p. 307, n. 16.)

115 Details of the discussions in London in Ullman, *Intervention and the War*, p. 307.

116 From Eastern Committee Minutes, 28 May and 31 May 1918, given in Ullman, *Intervention and War*, p. 307.

117 Lloyd George in the War Cabinet, 24 June 1918; London, PRO; War Cab. 435, Min. 8, CAB 23/6.

118 Mints, *God 1918*, pp. 470–3; Mikoyan, *Memoirs of Anastas Mikoyan*, I, pp. 135–7.

119 Goldsmith's report to the Director of Military Intelligence, 1 July 1919; London, PRO; WO 95/4960.

120 MacDonell, *And Nothing Long*, pp. 210–2, 251.

121 Dunsterville, *The Adventures of Dunsterforce*, p. 170.

122 Rawlinson, *Adventures in the Near East*, p. 65. Goldsmith provides some extra information about Rawlinson's movements, according to which Rawlinson was in the Caucasus since early 1918 on a secret mission, disguised as Mr Robinson, an American engineer. There is no reference to this particular mission in Rawlinson's own account. (Goldsmith's report to the Director of Military Intelligence, 1 July 1919; London, PRO; WO 95/4960.)

123 See Mitrokhin, *Failure of Three Missions*, pp. 39–40; Dunsterville, *The Adventures of Dunsterforce*, p. 208; Brinkley, *The Volunteer Army and Allied Intervention in South Russia, 1917–1921*, p. 63.

124 Rawlinson, *Adventures in the Near East*, pp. 64–7; Dunsterville, *Adventures of Dunsterforce*.

125 Given in Mitrokhin, *Failure of Three Missions*, p. 45.

126 Moberly, *History of the Great War*, p. 183; Kazemzadeh, *The Struggle for*

Transcaucasia, pp. 133–4.

127 Anastas Ovanesovich Mikoyan states in his memoirs that 'to be on the safe side, it was decided not to have him [Bicherakhov] enter Baku, but to have him halt at the Aliat [Alyat] Station …'(*Memoirs of Anastas Mikoyan*, I, pp. 120–1.)

128 Dunsterville, *The Adventures of Dunsterforce*, pp. 196–7. Major–General Macdonough, director of British Military Intelligence, however, seems to have been misinformed about Bicherakhov's movements. Both on 23 July and 20 August he informed the Imperial War Cabinet that 'Bicherakhov had never gone to Baku'. (Imperial War Cabinet Minutes; London, PRO; Cab. 23/41 and Cab. 23/42.)

129 *Izvestiia* (Baku), 9 July 1918, in Suny, *The Baku Commune, 1917–1918*, p. 291.

130 From Senior Naval Officer(Baghdad) to Admiralty, 10 November 1918; London, PRO; FO 371/3347. Mints, *God 1918*, p. 471. Mikoyan, *Memoirs of Anastas Mikoyan*, I, p. 121.

131 Rüstü, 'Büyük Harpte Kafkas Yollarinda 5–nci Piyade Firkasi', *Askeri mecmua*, No. 92, 1934.

132 Hovannisian, *Armenia on the Road of Independence*, pp. 204–5.

133 Stalin, 21 July 1918, *DVP*, I, pp. 401–2. All this correspondence is summarised in the letter from Shaumian to Lenin, 31 July 1918; Moscow, TsPA; Fond: 160, Op.: 3, D.: 6.

134 Azizbekova, Mnatsakanyan and Traskunov, *Sovetskaia Rossiia i Borba za Ustanovlenie i Uprochenie Vlasti Sovetov v Zakavkaze*, p. 97.

135 Given in Suny, *Baku Commune*, p. 316.

136 Telegramme from Shaumian(Baku) to Lenin, 27 July 1918, *DVP*, I, pp. 411–12.

137 Given by Aralov, *Vospominaniia Sovetskogo Diplomata*, p. 23. There is a Turkish translation of this book under the title of *Bir Sovyet Diplomatinin Türkiye Hatiralari*.

138 Shaumian and 25 other Baku commissars were arrested soon after by an anti–Bolshevik committee based in Ashkabad, and later were taken out into the desert and murdered on 20 September 1918. (for a full list of the commissars see Mikoyan, *Dorogoi Borbi*, pp. 226–42.) When the news of the massacre reached Moscow in early 1919 the Bolshevik government accused British officers of being responsible for the death of the commissars . (Azizbekova, Mnatsakanyan, Traskunov, *Sovetskaia Rossiia i Borba za Ustanovlenie i Uprochenie Vlasti Sovetov v Zakavkaze*, p. 103; Mitrokhin, *Failure of Three Missions*, p. 49.) The British government, however, denied any role in the incident. *(Command Paper 1846*, Russia No. 1, 1923.)

139 The Centro–Caspian Directorate is described as a 'tool of the British imperialism' in Soviet accounts. (Azizbekova, Mnatsakanyan, Traskunov, *Sovetskaia Rossiia i Borba za Ustanovlenie i Uprochenie Vlasti Sovetov v Zakavkaze, p.* 99; Mitrokhin, *Failure of Three Missions*, p. 46.)

140 Dunsterville, *The Adventures of Dunsterforce*, p. 284; Moberly, *History of the Great War*, pp. 202–3.

141 ibid., pp. 198–9.

142 ibid., p. 213.

143 Rawlinson, *Adventures in the Near East*, pp. 67–9.

144 Allen and Muratoff, p. 492; Dunsterville, *The Adventures of Dunsterforce*, pp. 230, 284; Kazemzadeh, *Struggle for Transcaucasia*, p. 139; Moberly, *History of the Great War*, pp. 105, 216, 246; Mitrokhin, *Failure of Three Missions*, p. 46; Hovannisian, *Armenia on the Road to Independence 1918*, p. 222.

145 Moberly writes that the Turks opposite Baku numbered 5,300 regulars with 26 guns and 8,000 Tatar irregulars under Turkish officers. (Moberly, *History of the Great War*, p. 230.)

146 Dunsterville, *The Adventures of Dunsterforce*, p. 309.

147 Russian–German Supplementary Treaty to the Peace Agreement between Russia, and Germany, Austro–Hungary, Bulgaria and Turkey, 27 August 1918, *DVP*, I, pp. 443–5; an English account of this treaty in Wheeler–Bennet, *Brest–Litovsk*, p. 433;

148 A radio telegram sent by Ludendorff on 28 June was deciphered by the Caucasus Military Agency, in Goldsmith's report; London, PRO; WO 95/4960.

149 ibid.; Yerasimos, *Türk–Sovyet Iliskileri*, pp. 30–1.

150 Dunsterville, *The Adventures of Dunsterforce*, p. 275.

151 ibid., p. 279.

152 Moberly, *History of the Great War*, p. 230.

153 Dunsterville, *The Adventures of the Dunsterforce*, pp. 291–2 and Rawlinson, *Adventures in the Near East*, p. 83.

154 MacDonell, *And Nothing Long*, p. 265. The Caucasus Military Agency, on the other hand, had been transferred to Vladikavkaz when the Turkish armies approached the Tiflis area. There Colonel Pike was accidentally shot on 15 August while watching the street battles between the Bolsheviks and the Terek Cossacks. Goldsmith, who after the death of Pike became the Acting Commanding Officer of the Caucasus Military Agency, and the other members of the mission were arrested in Vladikavkaz by the Bolsheviks in October for being connected to Lockhart conspiracy. Lockhart was then under arrest in Moscow for engineering a counter–revolutionary plot and was eventually sentenced to death. ('Secret and Confidential Memorandum on the Alleged "Allied Conspiracy" in Russia', 5 November 1918; London, PRO; FO 371/3348.) Members of the Caucasus Military Agency were sent to Moscow in January 1919 and stayed in Butirski prison until July 1919. (Goldsmith's Report to the Director of Military Intelligence, 1 July 1919; London, PRO; WO 95/4960.)

155 Branch Memoranda: South Russia, Caucasus 1918 August–1919 March, by Admiralty Naval Staff Operations Division; London, PRO; WO 106/1206.

156 Azizbekova, *Sovetskaia Rossiia i Borba za Ustanovlenie i Uprochenie Vlasti Sovetov v Zakavkaze*, pp. 101–2. Following the fall of Baku to the Turks the revenge for the March Days in the Armenian sections of the city was ferocious. Conservative estimates put the number of lives lost between 9,000

and 10,000. (Swietochowski, *Russia and Azerbaijan*, p. 73.)

157 Dunsterville, *The Adventures of the Dunsterforce*, p. 261; Rawlinson, *Adventures in the Near East*, pp. 73, 80; MacDonell, *And Nothing Long*, p. 279.

158 Given by Arslanian, 'Dunsterville's Adventures: A Reappraisal', *International Journal of Middle Eastern Studies*, XXII/2, p. 209.

159 Teague–Jones was accused by the Bolshevik government of being personally responsible for the deaths of the 26 Baku commissars. (Mints, *God 1918*, p. 474.) Preferring not to risk Moscow's vengeance, Teague–Jones changed his name and disappeared from view. He spend the rest of his life until his death in 1988 behind the false identity of Ronald Sinclair. (Peter Hopkirk's introduction to Teague–Jones, *The Spy Who Disappeared*, pp. 11–12.)

160 Diary of A Secret Mission to Russian Central Asia in 1918 by Reginald Teague–Jones, entrance 24.8.1918, in Teague–Jones, *The Spy Who Disappeared*, p. 101.

161 Ullman,*Intervention and the War*, p. 330.

162 In January 1831 a British officer in disguise, 24 year old Lieutenant Arthur Conolly was 'sent into field to reconnoitre the military and political no-man's-land between the Caucasus and the Khyber, through which a Russian army might march'. He was the 'archetypal Great Game player, and it was he, fittingly enough, who first coined this memorable phrase in a letter to a friend'. (Hopkirk, *The Great Game*, p. 123.)

Chapter 2

1 Temperley, *A History of the Peace Conference of Paris*, VI, p. 55.

2 Imperial War Cabinet Minutes; London, PRO; Cab. 23/42.

3 'Armistice Convention with Turkey, 30 October 1918', and 'Armistice concluded by Great Britain and Allied and Associated Powers with Germany, 11 November 1918', in *British and Foreign State Papers*, 1917–1918, CXI, pp. 611–24.

4 Lloyd George, *War Memoirs*, II, pp. 1946–7.

5 Kutay, *Atatürk-Enver Pasa Hadiseleri*, p. 15; Tunaya, *Türkiye'de Siyasi Partiler, 1859–1952*, p. 180.

6 Ahmet Izzet Pasha(1864–1937). Chief of the general staff after 1908. War minister in 1913. During the First World War he served on the Caucasian front. In 1918 succeeded Talat as grand vizier. (Gövsa, *Türk Meshurlari Ansiklopedisi*, pp. 199–200.)

7 On 23 January 1913 Enver and a small party of officers had launched an armed assault on the Sublime Porte. They had shot the anti-CUP minister of war and forced the grand vizier to write his resignation at the point of their guns. From then on the CUP single-handedly run the country by a virtual military dictatorship until the end of the war. (For the CUP see Ahmad, *The Young Turks. The Committee of Union and Progress in Turkish Politics 1908–14*; Aksin, *Jön Türkler ve Ittihat Terakki*; Arai, *Turkish Nationalism in the Young Turk Era*; Heyd, *Turkish Nationalism and Western Civilisation.*)

8 Hüseyin Rauf (Orbay) (1881–1964). He won fame as the captain of the cruiser *Hamidiye* during the Balkan Wars of 1912–1913. Served in the navy and as Ottoman agent in Persia during the First World War. (Gövsa, *Türk Meshurlari Ansiklopedisi*, pp. 292–3.)

9 Armistice Documents; London, PRO; FO 371/3449/181110. The general conditions of the Armistice were decided earlier at the 'Meetings of the Conference of British, French and Italian Prime Ministers' between 5 and 7 October. (Imperial War Cabinet Minutes, 11 October 1918; London, PRO; Cab. 23/42.)

10 Orbay, 'Rauf Orbay'in Hatiralari', *Yakin Tarihimiz*, II, pp. 48–9.

11 For an investigation of the prelude to the armistice negotiations, see Dyer, 'The Turkish Armistice of 1918, 2. Lost Opportunity : the Armistice Negotiations of Moudros', *Middle Eastern Studies*, VIII(1972), pp. 313–48.

12 On the armistice see Türkgeldi, *Mondros ve Mudanya Mütarekelerinin Tarihi*, pp. 23–73. This is an important account written by the secretary of the Turkish delegation at the armistice negotiations in Mudros.

13 The rough minutes of the negotiations at Mudros are in London, PRO; FO 371/5259/E 5732/5732/44.

14 *British and Foreign State Papers*, (1917–1918), V, CXI, pp. 611–13; Temperley, *History of the Peace Conference of Paris*, I, p. 495.

15 The minutes of the negotiating sessions at Mudros, London, PRO, FO 371/5259,E 5732/5732/44.

16 Italian Foreign Minister Baron Sidney Sonnino at the first formal session of the Allied Supreme War Council which met at Versailles on the morning of 31 October. (Given in Rudin, *Armistice 1918*, p. 294.)

17 'Outline of the Events in Transcaucasia from 1917 to 1921' by W. J. Childs and A. E. R. McDonnell, 31 May 1922, p. 10; London, PRO; FO 371/7729/E 8378.

18 The other members of the British High Commission were Rear Admiral Richard Webb as assistant high commissioner, J. B. Hohler as Calthorpe's principal advisor on the political side, Sir Adam Block as the financial expert, and Brigadier-General Deedes as the military attaché. (Clayton, *British Empire*, p. 64; Mears, *Modern Turkey*, p. 555.)

19 To set up a separate command under General Milne to control Turkey and the Caucasus was first proposed by Lloyd George on 3 December 1918 and accepted by the representatives of the other Allies. ('Notes of an Allied Conversation held on 3 December 1918'; London, PRO; Cab. 23/42.)

20 Moberly, *History of the Great War Based on Official Documents, The Campaign in Mesopotamia 1914–1918*, IV, pp. 329–30.

21 According to the records of the Ottoman Ministry of the Interior the population of Constantinople at the beginning of the war totalled 909,978 of which 560,434 ere Muslim, 205,375 were Greek, 72,962 Armenian and 52,126 were Jews. (Karpat, *Ottoman Population, 1830–1914*, pp. 170–1.) These figures were close to those prepared by General Milne in November 1919: total: 848,000, with 560,000 Turks, 205,000 Greeks and 83,000 Armenians. Milne does not give a figure for city's Jewish population. (Cambridge, Trinity

College Library; Montagu Papers, AS 4/7/19(2).)

22 Price, *The Rebirth of Turkey*, p. 126.

23 Karabekir, *Istiklal Harbimiz*, pp. 5–16. See also Chicherin's statement on the measures taken by the Allied troops in Asia Minor, 20 April 1919, Moscow, AVP, Fond: Reference about Turkey, Op.: 2, Por.: 2, Pap.: 2.

24 Nicolson, *George Curzon: The Last Phase, 1919–1925*, p. 63.

25 Armistice between the Allies and Germany was signed on 11 November 1918. (An English translation of the Armistice Agreement in *British and Foreign State Papers*, 1917–1918, CXI, pp. 611–24; and Barclay, *Armistice 1918*, pp. 132–45.)

26 For a discussion of the origins of the divisions among the Allies see Mears, *Modern Turkey*, pp. 555–6, and Toynbee, *The Western Question in Greece and Turkey*.

27 Lloyd George, *War Memoirs*, VI, pp. 3309–10.

28 Gilbert, *Winston S. Churchill*, IV, pp. 167–80.

29 Hurewitz, *Diplomacy in the Near and Middle East*, II, pp. 7–25. For a brief Soviet account of these diplomatic arrangements see Kriazhin's article 'Razdel Turtsii vo Vremia Mirovoi Voiny' which was published in *Novyi Vostok* in 1923. (no. 4, pp. 49–57.)

30 Miller, *My Diary at the Conference at Paris with Documents*, III, pp. 303–7; Lloyd George, *Memoirs of the Peace Conference*, II, p. 655.

31 On 30 December 1918 Lloyd George informed the Imperial War Cabinet that President Wilson expressed himself in favour of the Turks being cleared out of Europe altogether. (Imperial War Cabinet Minutes; London, PRO; Cab. 23/42.)

32 From Montagu to Hankey, 2 October 1918; Cambridge, Trinity College Library; Montagu Papers, Montagu AS 4/6/1–25.

33 On October 3, 1918, Hankey replied Montagu by saying that 'The line which you urge in your letter of 2nd October about Turkey is exactly the view the War Cabinet take'. From Hankey to Montagu, 3 October 1918; Cambridge, Trinity College Library; Montagu Papers, Montagu AS 4/6/2.

34 The armistice convention with Bulgaria, signed on 29 September 1918, allowed the Bulgarian administration to continue to exercise its functions on Bulgarian lands, including those at the time of the armistice occupied by the Allies. (Rudin, *Armistice 1918*, pp. 404–5.)

35 Ronaldshay, *The Life of Lord Curzon*, III, p. 260.

36 Nicolson, *George Curzon: The Last Phase 1919–1925*, pp. 76–8.

37 The Paris Peace Conference was convened on 12 January 1919. For a detailed study of the conference see Mayer, *Politics and Diplomacy of Peacemaking: Containment and Counterrevolution at Versailles, 1918–1919*.

38 Demobilisation started immediately after the end of the war. On 2 December 1918 Lieutenant-General Macdonough, this time holding the post of adjutant general to the forces, proposed to the Imperial War Cabinet the demobilisation of the army to a force of 21 divisions with the accessories, exclusive of the troops required in the UK, India and other garrisons abroad. This was agreed to. (Imperial war Cabinet Minutes,; London, PRO; Cab. 23/

42.) In early January 1919, there were demonstrations by soldiers at Dover and Folkestone demanding immediate demobilisation. By the end of January almost a million officers and men had already been demobilised. (Gilbert, *Winston S. Churchill*, IV, pp. 181–93.)

39 Churchill, *The World Crisis, Part IV, The Aftermath*, p. 166; Churchill, *The Stricken World*, IV, p. 472.

40 Churchill, *The World Crisis, Part IV, The Aftermath*, p. 415.

41 Eleutherios K. Venizelos (1864–1936) Founder and leader of the Greek Liberal party from which derive the modern parties of the centre. Friend of Lloyd George and admirer of British parliamentary institutions. (Alantos, D. *Venizelos*, London: Humpries, 1942)

42 To support the Greek position at the Peace Conference Greek clubs and benefactors passed resolutions, sponsored lectures, and published articles in European capitals. The Greek national anthem was published in the *Daily Telegraph*, which was translated into English by Rudyard Kipling. (Smith, *Ionian Vision*, pp. 63–4.)

43 For the Greek claims see Venizelos, *Greece Before The Peace Congress of 1919*. This is a long formal statement of Venizelos, written in Greek and published in English and French.

44 The river Maritza divides Thrace into two distinct sections, namely western Thrace and eastern Thrace.

45 'Weekly Intelligence Summary for the week ending March 8, 1919. No. 93, *United States Military Intelligence*, VIII, p. 653; *Reply of the Hellenic Delegation*, pp. 40–2.

46 Smyrna[modern Izmir] was one of the few ports in Anatolia, where the value of the exports exceeded that of the imports, a condition of affairs largely due to the Smyrna–Aydin railway, which was built in 1856 by the British. Besides increasing the export trade of Smyrna, the railway gave an immense stimulus to the domestic commerce of the interior.

47 Venizelos, *Greece Before The Peace Congress of 1919*, pp. 21–2.

48 ibid., pp. 18–19.

49 Diomidis-Petsalis, *Greece at the Paris Peace Conference*, p. 102.

50 Diomidis-Petsalis suggests that even if Greece had been offered the Turkish capital, it would have declined. This view is supported by the claim that '… their [the Entente] motive would have been to avoid the burden of having to control the Straits and to be in a position to refuse other Greek claims.'(Diomidis-Petsalis, *Greece at the Paris Peace Conference*, p. 102.) Venizelos's policy also might be explained in the light of an alleged plan of promoting Greco–Turkish co-operation. A Turkish person named Basri Bey, it is claimed, met with Venizelos that he would use his influence with the Sultan to promote Greco–Turkish co-operation. Basri Bey's proposal was based upon the belief that Turkey would suffer less under a Greek influenced regime than any other regime in which foreigners had a voice. (Diomidis-Petsalis, Greec*e at the Paris Peace Conference*, p. 190). According to this view, if Constantinople remained Turkish, then that city might become the home for a Greco–Turkish confederation. (ibid.)

51 Hurewitz, *Diplomacy in the Near and Middle East* . II, p. 11.

52 The major decision-making bodies of the Paris Peace Conference were the Supreme War Council, a group carried over from the war; the Council of Ten, a body based on the Supreme War Council but without military advisors and consisting of two plenipotentiary representatives – usually the head of government and the foreign minister – from each of the major powers, England, France, Italy, Japan, and the United States; and the Council of Four, Clemenceau, Lloyd George, Orlando and Wilson.

53 Nicolson, *Peacemaking, 1919*, p. 280. This is an important diary written by a technical adviser to the Greek Committee and an advisor to the Big Four on the question of the division of European and Anatolian Turkey. After listening to Venizelos at the conference, Nicolson wrote to his father that 'He[Venizelos] and Lenin are the only two really great men in Europe'. (ibid., p. 221.)

54 Weekly Intelligence Summary, for the week ending May 17, 1919, No. 103, in *United States Military Intelligence*, VIII, p. 954.

55 Hurewitz, *Diplomacy in the Near and Middle East*, II, p. 37.

56 Baker, *Woodrow Wilson and the World Settlement*, II, p. 191.

57 'Record of attitude since the outbreak of war in 1914', by Mr Adams, October 8, 1922, London, PRO, FO 371/E 10728/27/44, 7900.

58 A letter to Montagu from Joseph Bliss (who was in the city during the landing) describes the landing in detail, June 16, 1919, Cambridge, Trinity College Library, Montagu Papers, AS 4/4/31–42.

59 Ariburnu, *Milli Mücadelede Istanbul Mitingleri*, p. 212; Cebesoy, *Milli Mücadele Hatiralari*, pp. 58–61.

60 'Report of the Interallied Commission of Enquiry', *British Documents*, II, No. 17, Appendix A. Balfour to Curzon, 18 June 1919; London, PRO; FO 406/41; Toynbee, *The Western Question in Greece and Turkey*, pp. 270–3, 390–405.

61 Churchill, *The World Crisis, Part IV, The Aftermath*, p. 388–90; Nicolson, *George Curzon: The Last Phase, 1919–1925*, p. 94.

62 In a War Office statement, dated 12 December 1921, the Greek landing in Smyrna is described as the crucial event which started both the Turkish resistance movement and the close co-operation of the Turks with the Bolsheviks against the British. (From War Office, 12 December 1921, London, PRO, FO 371/ E 13700/143/44, 6537.)

63 Biyiklioglu, *Trakya'da Milli Mücadele*, p. 129; Novichev, 'Antikrest'yanskaia politika Kemalistov v 1919–1922', *Voprosy Istorii*, 9(1951), p. 64.

64 Tanör, *Türkiye'de Yerel Kongre Iktidarlari*, pp. 18–44.

65 For contemporary Turkish interpretations of the fourteen points of Wilson see Criss, *Isgal Altinda Istanbul*, pp. 84–7.

66 *The Papers of Woodrow Wilson*, XXXXV, p. 528. Lloyd George, in his memoirs, writes that 'we never formally accepted them [the fourteen points of Wilson], and they constituted no part of the official policy of the Alliance.' *War Memoirs*, V, p. 2489.

67 Adivar, *The Turkish Ordeal*, p. 36; Cebesoy, *Milli Mücadele Hatiralari*, pp.

55–60; Ariburnu, *Milli Mücadele'de Istanbul Mitingleri.* According to a number of accounts these public demonstrations were organised by the *Teskilat-i Mahsusa* (Mission Special), the intelligence organisation of the CUP. *Teskilat-i Mahsusa* had been founded before the war as the secret political police of the Ottoman state. After the Mudros Armistice most of its intelligence network remained intact and worked together with the nationalist movement. (Criss, *Isgal Altinda Istanbul*, pp. 143–9; Zürcher, *The Unionist Factor*, p. 59.)

68 From Zaydel(Odessa) to Chicherin(Moscow), 20 April 1919; Moscow, AVP; Fond: Reference about Turkey, op.: 2, por.: 1, pap.: 2. Later Soviet accounts, however, describe the emergence of Turkish resistance in Anatolia as a direct consequence of the Bolshevik Revolution in Russia. (Rozaliev, *V. I. Lenin i Turtsiia*, p. 184.)

69 The first detailed Soviet account on the rise of Turkish national movement and Mustafa Kemal appeared in 1921. (Pavlovitch [Veltman], 'Kemalisticheskaia dvizhenie v Turtsii', *Krasnaia Nov*, 1, pp. 218–22.)

70 A short account by Mustafa Kemal of his role in the Dardanelles campaign was published by Igdemir, *Anafartalar Muhaberatina Ait Tarihçe.*

71 Rauf, *The Last Sultans*, pp. 350–1; Ozerdim, *Atatürk Devrimi Kronolojisi*, p. 12. (A Turkish law of 1934 obliged all citizens to take family names, which few Turks heretofore possessed. Mustafa Kemal was given the surname 'Atatürk', meaning the father of the Turks, by the GNA.)

72 Price, *Extra-Special Correspondent*, p. 104.

73 Mustafa Kemal delivered a famous Six-Day Speech in 1927 which in turn was transformed an essential text material for the official Turkish historians. Kemal used this public speech to justify all his political decisions and actions and to answer some of the criticism raised by his opponents. (Atatürk, *Nutuk*. An English version of it was published under the title of *A Speech Delivered by Mustafa Kemal Atatürk. 1927*.)

74 Atatürk, *Nutuk*, pp. 1–4.

75 Cebesoy, *Milli Mücadele Hatiralari*, p. 37.

76 Karabekir, *Istiklal Harbimiz*, pp. 12–16.

77 Karabekir, *Istiklal Harbimiz*, pp. 14–15. Biyiklioglu, *Atatürk Anadolu'da*, p. 11.

78 Izzet Pasha explained the reason to why he did not include Mustafa Kemal in his cabinet by saying that Kemal was too demanding and irascible. (In Aydemir, *Tek Adam*, I, p. 334)

79 Atatürk, *A Speech Delivered by Mustafa Kemal Atatürk. 1927*, p. 7.

80 Jaschke, *Türk Kurtulus Savasi Kronolojisi*, I, p. 24.

81 Summary of Economic and Political Situation in Turkey, by the Executive Committee of the Comintern Eastern Department, 1 January 1922; Moscow, TsPA; Fond: 5, Op.: 3, D.: 630.

82 Mehmet VI Vahdettin (1861– 1929). Thirty-sixth and last Ottoman Sultan. Ruled from 3 July 1918 to 17 October 1922. Sultan Mehmet V died in July 1918 and was succeeded by his brother Vahdettin, who ascended the throne under the name of Mehmet VI. Vahdettin was deposed as a result of his pro–British stand in the years after World War I. Thereafter he lived

in exile on Malta, in Genoa, and in San Remo where he died. (Karal, M*ehmed VI. Muhammed VI Vahid al-Din*, IA, VII, pp. 562–66.)

83 Orbay, 'Rauf Orbay'in Hatiralari', p. 402; Jaschke, *Kronoloji*, I, p. 7; Rauf, *The Last Sultans*, p. 353; Mansel, *Constantinople*, p. 388.

84 The orthodox Turkish historiography, which is ultimately based on Mustafa Kemal's own words in his *Nutuk* presents Mustafa Kemal's arrival in Samsun as the start of the national resistance. (Karal's *Türkiye Cumhuriyeti Tarihi 1919–1944*, Kili's *Türk Devrim Tarihi*, and *Tarih. Dördüncü Cilt. Türkiye Cumhuriyeti* (prepared by Turkish Historical Association) represent the mainstream of orthodox Kemalist historiography.)

85 From the Diaries of Sir Maurice Hankey (secretary to the cabinet), 16 October 1919; Cambridge, Churchill College Archive Centre; HNKY, 1/5, p. 94.

86 On 23 October 1919, the British Intelligence Staff in Constantinople was reporting that the Turkish population of the capital was ready 'to resist any attempt to occupy the country'. (General Staff Intelligence Report, British High Commission, Constantinople, 23 October 1919; London, PRO; FO 371/4160/E 149600.)

87 Nicolson, *George Curzon: The Last Phase, 1919–1925*, p. 69.

88 Curzon at the Imperial War Cabinet, 23 December 1918; London, PRO; Cab. 23/42.

89 'Outline of Events in Transcaucasia from the Beginning of the Russian Revolution in the Summer of 1917 to April 1921', by W. J. Childs and A. E. R. McDonnell, 31 May 1922; London, PRO; E 8378/8378/58 FO 371/6280.

90 The Memorandum, by Major-General P. de B. Redcliffe, Director of Military Operations, 13 November 1918; London, PRO; Cabinet Paper G. T. 6274, Cab. 24/69.

91 Raevskii, *Angliiskaia Interventsiia*, pp. 108–9.

92 Hovannisian, *The Republic of Armenia*, pp. 78–92.

93 Denikine, *The White Army*, p. 156.

94 From the War Office to the G.H.Q. (Constantinople), 10 March 1919; London, PRO; WO 157/767.

95 Imperial War Cabinet Minutes, 20,25,27,28 June and 23 July 1918; London, PRO; Cab. 23/41.

96 'Memorandum on Siberia', FO Intelligence Bureau, Dept. of Information, 14 March 1918; London, PRO; FO 371/3290. From Lockhart (Moscow) to the FO, 6 April 1918; London, PRO; FO 371/3290. Lord Milner (secretary of state for war) estimated, in December 1918, the number of British troops at about 13,000 in north Russia and at 4,000 in Siberia. There were also some 10,000 Americans, some 60,000 Japanese, and a few thousand French and Italian troops. (Imperial War Cabinet Minutes, 23 December 1918; London, PRO; Cab. 23/42.)

97 Foreign Office Minute by Balfour, 29 November 1918; London, PRO; FO 371/3345.

98 From Curzon to Balfour (Paris), FO 11 June 1919, *British Documents*, III, p. 365, and for the terms of the convention ibid., pp. 369–70.

99 Memorandum by Fisher, 5 December 1918; London, PRO; Cabinet Paper GT 6443, CAB/24/71.

100 Imperial War Cabinet Minutes, 23 December 1918; London, PRO; Cab. 23/42.

101 In the second volume of his three volume work on Anglo–Soviet relations, Ullman provides a detailed account of the discussions on the Caucasus in the British Cabinet and Eastern Committee. *Britain and the Russian Civil War*, pp. 64–98.

102 Imperial War Cabinet Minutes, 12, 23, 30 December 1918; London, PRO; Cab. 23/42.

103 Alexander Kolchak established his government in Omsk on 18 November 1918. (Brinkley, *The Volunteer Army and Allied Intervention in South Russia*, p. 59.)

104 After its retreat from the Don region in January 1918, the Volunteer Army undertook operations in the Kuban region, and Anton Ivanovich Denikin undertook the command of it when Kornilov was killed in April 1918. (ibid., p. 23.)

105 Enclosure ('Proclamation of Tchaikovski, the head of the North Russian Government') to the telegram from Mr Lindley(Archangel) to the FO, 13 November 1918; London, PRO; FO 371/3348.

106 Telegram to Paris, Rome, Washington, and Tokyo embassies, 2 January 1919; London, PRO; file 1347/91/38, FO 371/3954. The resulting memorandum to the State Department, 3 January, is in Department of State, *Foreign Relations of the United States, 1919, Russia*, pp. 2–3.

107 'Secretary's notes of a conversation held in M. Pichon's room at the Quai d'Orsay', 16 January 1919, *Foreign Relations of the United States: The Paris Peace Conference, 1919*, III, pp. 578–84.

108 For a Soviet account of the 'Russian Question' in the peace conference see Stein, '"Russkii Vopros" no Parizhskoi Mirnoi Konferentsii', *Voprosy Istorii*, 8(1947), pp. 3–29.

109 British Empire Delegation, Paris, 2nd minutes, 20 January 1919, 4 p.m.; London, PRO; Cab. 29/28/1.

110 ibid.

111 Telegram from Osborne, Charge in Denmark, to the Commission to negotiate peace(Paris), 18 January 1919, *Foreign Relations of the United States, 1919, Russia* , pp. 15–17; Report, dated 18 January 1919, of agent who held confidential conversations with Litvinoff on 14, 15, and 16 January, *Foreign Relations of the United States: The Paris Peace Conference, 1919*, III, pp. 643–6; Temperley, *A History of Peace Conference in Paris*, VI, p. 312.

112 Statement drafted by President Wilson to be issued to the warring factions in Russia, 22 June 1919, *Foreign Relations of the United States: The Paris Peace Conference, 1919*, III, p. 691.

113 Secretary's notes of a conversation held in M. Pichon's room at the Quai d'Orsay, Paris, 22 January 1919, *Foreign Relations of the United States, 1919, Russia*, pp. 30–1.

114 There exist some errors on the name and the location of Prinkipo

where the proposed meeting would be held. R. Ullman, for instance, mentions Prinkipo as the 'local name for the Princes Islands.'(*Britain and the Russian Civil War*, p. 109.); F. S. Northedge and Audrey Wells refer to Oxya [*Sivri Ada*] as the island chosen. Oxya's chief fame was that on several occasions all the wild dogs of Constantinople were rounded up and exiled there where they soon ate each other. *(Britain and Soviet Communism*, p. 30.) Andrew Rothstein, on the other hand, writes that Prinkipo is 'an island in the Aegean'..*(A History of the USSR*, p. 125.) For more information on the Princes Islands see Sumner-Boyd and Freely's excellent reference book, *Strolling Through Istanbul*, pp. 432–7.

115 *Atatürk'ün Tamim Telgraf ve Beyannameleri, 1917–1938*, p. 126.

116 *DVP*, II, pp. 45–6.

117 Vaclav Vaclavovich Vorovsky joined the Party in 1894. After the Bolshevik Revolution he became the Soviet state's plenipotentiary in Scandinavia (1917–19). He was appointed as the head of the State Publishing House (1919–20), and then sent to diplomatic work in Italy (1921–23). (Piyashev, *V. V. Vorovsky*; Senn, *Assassination in Switzerland, The Murder of Vatslav Vorovsky*.)

118 Telegram from Chicherin to Vorovsky(Sweden), 20 January 1919, *DVP*, II, pp. 42–5.

119 Telegram from Lenin to Trotsky(Kozlov), 24 January 1919, Meijer, *The Trotsky Papers*, I, pp. 259–61.

120 ibid., p. 261.

121 His fate, however, brought him to Prinkipo, ten years later. This small, peaceful island became Trotsky's first stop in his exile years when Stalin expelled him from the Soviet Union in 1929.

122 Radio message from Chicherin to Wilson (Paris), 28 January 1919, *DVP*, II, p. 52.

123 Minutes, Council of Ten, 1 February 1919, 3 p.m., *Foreign Relations of the United States: The Paris Peace Conference, 1919* , III, pp. 835–6.

124 The message from the People's Commissariat for Foreign Affairs to the American representative, Mr Poole, on 5 August 1918; to President Wilson on 24 October 1918; to all Entente governments through representatives of neutral countries, on 3 November 1918; a message from the Sixth All-Russia Congress of Soviets, on 7 November 1918; Litvinov's Note in Stockholm to all Entente representatives, on 23 December 1918; the messages of 12 and 17 January 1919. (From 'the draft resolution on foreign policy', written and presented by Lenin at the session of the Eight All-Russia Conference of the RCP(B), 2 December 1919, Lenin, *Collected Works*, XXX, p. 191.)

125 The Congress was held at the Bolshoi Theatre between 6 and 9 November 1918. Its opening coincided with the celebrations of the anniversary of the November Revolution. The agenda included the international situation and the military situation. On the proposal of Sverdlov, President of the Central Executive Committee, the Congress adopted an appeal to the governments at war in Russia to start peace negotiations. (ibid., XXVIII,

p. 508.)

126 Lenin's 'Speech on the International Situation', November 8, 1918, Extaordinary Sixth All-Russian Congress of Soviets, ibid., p. 154.

127 *Pravda*, 5 November 1918, p. 2; Lenin's 'Speech on the International Situation', 8 November 1918, Extaordinary Sixth All-Russian Congress of Soviets, *Collected Works*, XXVIII, pp. 151–64.

128 ibid.

129 The First Congress of the Communist International met in Moscow between 2 and 6 March 1919. The opening session on 2 March 1919 was attended by 52 delegates. Among the delegates were V. I. Lenin, V. V. Vorovsky and G. V. Chicherin. The Turkish communists were represented in the congress by Mustafa Subhi who presented a talk on the importance of the revolutionary work in the Eastern countries. (The First Congress of the Comintern is widely discussed in the third volume of Carr's *The Bolshevik Revolution*, pp. 123–33. For a Soviet account of the origins and the First Congress of the Comintern see *Kommunisticheskii Internatsional*, pp. 19–99.)

130 Degras, *Communist International*, I, p. 30.

131 Lenin's speeches at the First Congress of the Comintern in *Sochineniia*, XXXVII, pp. 487–511; *Collected Works*, XXVIII, pp. 453–77.

132 Note of the Soviet government to the government of Great Britain, France, Italy, the United States of America and Japan, 4 February 1919, *DVP*, II, pp. 57–60.

133 The Georgian Delegation at Paris to the Commission to negotiate peace, 8 February 1919, *Foreign Relations of the United States, 1919, Russia*, pp. 47–50.

134 Denikin, *Ocherki russkoi smuty*, II, p. 342.

135 Temperley, vol. VI, p. 313; Lloyd George, *Memoirs of the Peace Conference*, I, p. 240.

136 Strakhovsky, *Intervention at Archangel*, pp. 144–5.

137 A conference of Russian ambassadors and representatives was held in Paris in January 1919, and a representative body, 'the Russian Political Conference', was organised. (Brinkley, *The Volunteer Army and Allied Intervention in South Russia*, p. 104.)

138 Churchill, *The World Crisis, Part IV, The Aftermath*, p. 170.

139 Cited in Thompson, *Russia, Bolshevism and the Versailles Peace*, pp. 122–3.

140 ibid.

141 Churchill, *The World Crisis, Part IV, The Aftermath*, pp. 172–5; Lloyd George, *The Truth About the Peace Treaties*, I, pp. 370–1; Riddell, *Lord Riddell's Intimate Diary of the Peace Conference and After, 1918–1923*, p. 21.

142 According to Soviet accounts the whole Prinkipo episode represented a diplomatic success by the Bolshevik government which skilfully used the contradictions within the anti-Soviet camp. (See, for instance, Stein, 'Diplomatiia Antanty i proekt konferentsii na Printseviikti Ostrovakh', *Voprosy Istorii*, 3(1947), p. 52.)

143 William Allen White and Professor George Herron, the American

delegates to the Prinkipo Conference, claimed that 'officials of the French government' and 'the French agents in the East' told to the representatives of several Russian parties 'to have nothing to do with the conference...' (Briggs, *George D. Herron and the European Settlement*, p. 369.)

144 General accounts of the other initiatives include Ullman's *Britain and the Russian Civil War*, pp. 136–70; and Thompson's *Russia, Bolshevism and the Versailles Peace*, pp. 154–5.

145 Note for the Cabinet on 'Future Military Operations in Russia', 24 February 1919; London, PRO; Cab. 24/75, G.T. 6885.

146 Gilbert, *Winston S. Churchill*, IV, p. 265.

147 Weekly Intelligence Summary for the week ending March 15, 1919, No. 94, *United States Military Intelligence*, VII, p. 681; and the report by O.C. Harvey of the Russia Department, July 1919, *British Documents*, III, No. 342.

148 Calwell, *Field-Marshall Sir Henry Wilson: His Life and Diaries*, entry for 1 February 1919, pp. 164, 167–8.

149 Churchill, *Stricken World*, IV, p. 473.

150 On 16 March 1919 Lloyd George told the House of Commons that military intervention in Russia 'was the greatest act of stupidity that any government could possibly commit.'(*British Documents*, III, pp. 308–12.)

151 By the end of 1919 the number of British troops in Russia was reduced drastically from 40,000 to 2,000. (Given in Quinault, 'Churchill and Russia', *War and Society*, IX, 1(May 1991), p.103.)

152 Buchan, *The Baltic and Caucasian States*, pp. 219–20.

153 Hovannisian, *The Republic of Armenia*, pp. 479 and 482.

154 Ullman, *Britain and the Russian Civil War*, p. 358.

155 Busch, *Mudros to Lausanne: Britain's Frontier in West Asia, 1918–1923*, p. 161.

156 Calwell, *Field-Marshall Sir Henry Wilson*, II, pp. 167–8.

Chapter 3

1 Radio message from the Narkomindel, 27 September 1919; Moscow, AVP; Fond: Reference about Turkey, Op.: 2, Por.: 2, Pap.: 2.

2 London, PRO; FO 371/5046, E 3046/3/44, pp. 9–12. (In August 1919 Calthorpe handed over the high commissionership to Vice-Admiral Sir John M. de Robeck.)

3 From Mustafa Kemal(Ankara) to the Soviet Russian government, 26 April 1920; Moscow, AVP; Fond: Reference about Turkey, Op.: 3, Pap.: 2, D.: 3.

4 Lenin, *Collected Works*, XXIX, p.322.

5 Mikhail V. Frunze (1885–1925). Active in the RSDLP since 1904. During the Civil War he joined the Red Army and followed a military career. He fought on the Ural front, the Turkestan front, and in south Russia and in the Ukraine.

6 Speech in Orenburg(capital of the Orenburgskaia oblast in Russia), 20 September 1919, Frunze, *Neizvestnoe i zabytoe*, pp. 172–81.

7 *Pravda*, 9 November 1918.

8 Statement by Chicherin, 24 December 1920; Moscow, AVP; Fond: Reference about Turkey, Op.: 3, Por.: 2, Pap.: 2.

9 Narkomindel, 20 April 1919; Moscow, AVP; Fond: Reference about Turkey; Op:2; Por:2; Pap:2.

10 General Süleyman Sulkevich (1865–1920). A Lithuanian Muslim who had once served in the Russian army and during the war had commanded a special Muslim Corps for the German forces in Romania.

11 *Khronika, Revoliutsionnykh Sobytii v Krymu, 1917–1920 gg.*, pp. 67, 88, 92.

12 Bunyan, *Intervention, Civil War, and Communism in Russia*, p. 58.

13 Mustafa Subhi(1883–1921) was a Turkish journalist who had fled from exile in Sinop to Russia in 1914. Subhi participated in the First All–Russian Congress of Communist Organisations of Eastern People in November 1918 and became the head of the Turkish section in the Bureau of Eastern Nationalities. He had a consultative vote at the founding congress of the Comintern in March 1919. ('Theses about the Eastern and National Problem', Azerbaijan Communist Party, Baku, 20 February 1921; Moscow, TsPA; Fond: 5, Op.: 2, D.: 92. Undated 'Report to Lenin re. communist movements in Turkey', Mikhail Pavlovich; Moscow, TsPA; Fond: 5, Op.: 3, D.: 213.)

14 On 19 July 1918, for instance, before the conclusion of the war, the Ottoman representative in Moscow, Salih Kemal, had sent a letter of protest to Chicherin, regarding the anti–Ottoman socialist propaganda in the pages of *Yeni Dünya*, published in Moscow. Chicherin, in his reply to the Ottoman representative on the same day, had stated that the journal *Yeni Dünya* was an independent organ of Muslim socialists, and the Bolshevik government did not intend to control or ban their rightful activities. (Chicherin to Salih Kemal, 19 July 1918; Moscow, AVP; Fond: Reference about Turkey, Op:1, Por:3, Pap:1.)

15 *Izvestiia*, 23 April 1919.

16 *Mustafa Subhi Kavgasi ve Düsünceleri*, pp. 70–1.

17 Tansu, *Iki Devrin Perde Arkasi*, pp. 338–42; Yerasimos, *Turk Sovyet Iliskileri*, p. 108.

18 Karabekir, K. *Istiklal Harbimiz*, pp. 49–50; Yeraşimos, *Türk–Sovyet Iliskileri*, p. 109.

19 Karabekir, *Istiklal Harbimiz*, p. 47.

20 Refet Bele (1881–1963). Served with distinction on the Sinai front. Landed in Samsun with Mustafa Kemal as commander of the Third Army Corps.

21 'The Amasya Protocol', signed on 21 June 1919, became the first influential call for a nation-wide struggle against the Allied occupation. (Atatürk, *A Speech Delivered by Mustafa Kemal Atatürk. 1927*, pp. 205–10.)

22 Kandemir, *Mustafa Kemal, arkadaslari ve karsisindakiler*, pp. 67–8.

23 Orbay, 'Rauf Orbay'in Hatiralari', *Yakin Tarihimiz*, III, pp. 48–87.

24 In order to attract as much support as possible among a mixed population of Turks, Kurds, Arabs, Lazes and Circassians the nationalist leaders deliberately used the Arabic word '*millet*' for nation, implying a

religious community rather than an ethnic unity.

25 Dursunoglu, *Milli Mücadelede Erzurum*, p. 107; Atatürk, *A Speech Delivered by Mustafa Kemal Atatürk. 1927*, p. 57.

26 Halide Edib, a prominent writer, feminist, and supporter of the national movement, gives her account to why Kemal rose as the unquestionable leader of the national movement:

> He was by turns cynical, suspicious, unscrupulous, and satanically shrewd ... Of course, one knew all the time that there were men around him who were greatly his superior in intellect, moral backbone, and far above him in culture and education. But though he excelled them in neither refinement nor originality, not one of them could possibly cope with his vitality. Whatever their qualities, they were made on a more or less normal scale. In terms of vitality he wasn't. *(Turkish Ordeal*, p. 195.)

27 On 19 September 1919, just a week after the Sivas Congress, Damat Ferit, the prime minister of the Ottoman government in Constantinople, gave an interview to a French wire service. Damat Ferit asserted that Asia Minor was falling into the hands of the Bolshevik inspired groups. (Given by Paksoy, 'US and Bolshevik Relations with the TBMM Government: First Contacts. 1919–1921', *The Journal of Sophia Asian Studies*, XII(1994), pp. 218–19.)

28 On the night of 1 November Enver, Cemal, Talat and five others left aboard a German submarine for Odessa first. Enver apart, all of them were killed by Armenian assassins in 1920–1921.

29 Radek, Karl (Karl Sobelsohn) (1885–1939). He accompanied Trotsky to Brest–Litovsk. In November 1917 the Soviet government appointed him assistant commissar for foreign affairs. In April 1918 he was appointed head of Central European Department of Narkomindel. In this capacity he returned to Germany in 1918 and took part in the Spartacist revolt. After its failure he was imprisoned in the Moabit prison, in Berlin, from February 1919 to January 1920. During these eleven months he played the double role of adviser to the leader of the German CP, Paul Levi, and semi-official representative of the Bolshevik government to German politicians and military chiefs. After his release in December 1919 he returned to Russia and worked in the Comintern.

30 Tansu, *Iki Devrin Perde Arkasi*, p. 175. A brief interesting account of this episode in Carr, *German–Soviet Relations Between The Two World War Wars, 1919–1939*, p. 18 and 22–3.

31 Cebesoy, *Milli Mücadele Hatiralari*, p. 42; Sorgun, *Bitmeyen Savas. Kutulamare Kahramani Halil Pasanin Anilari*, pp. 320–1; Biyiklioglu, *Atatürk Anadolu'da*, p. 20.

32 Aydemir, *Makedonya'dan Orta Asya'ya Enver Pasha*, III, p. 586; Tansu, *Iki Devrin*, p. 177.

33 From Chicherin to the Central Committee, 18 April 1921; Moscow, TsPA; Fond: 5, Op.: 2, D.: 315.

34 Major Fritz Tschunke was later in the delegation to Soviet Russia, in

the spring of 1921, sent by *Sondergruppe R* (a special unit within the German Ministry of War specifically intended to conduct military negotiations with the Soviets). He later supervised the co-ordination of the production of German war materials in Soviet Russia. (Smith, 'The German General Staff and Russia, 1919–1926', *Soviet Studies*, 8(October 1956), pp. 125–32; Rosenbaum, *Community of Faith*, pp. 69, 288.)

35 Tschunke(Berlin) to Narkomindel, 14 July 1920; Moscow, AVP; Fond: Reference regarding Turkey, Op.: 3, Por.: 4, Pap.: 2.

36 Carr provides further evidence, quoting a letter from Enver to Seeckt on 26 August, 1920. The main theme of Enver's letter was the Soviet–German–Turkish co-operation. *(The Bolshevik Revolution*, III, p. 327.)

37 From Chicherin to Stalin, 14 September 1921; Moscow, TsPA; Fond: 5, Op.: 2, D.: 315.

38 Cebesoy, *Milli Mücadele Hatiralari*, p. 42; Müderrissoglu, *Kurtulus Savasinin Mali Kaynaklari*, pp. 543–5; Ozalp, *Milli Mücadele, 1919–1922*, I, p. 221.

39 I have not come across any direct evidence to whether the British knew about that particular agreement or not. However, the British documents of the period clearly show that there was a general suspicion about German support both to the Bolsheviks and the Turkish nationalists. In late 1919, Oliver Wardrop, the British representative in the Caucasus, mentioned in a telegram to Admiral de Robeck, the British high commissioner in Constantinople, that Denikin provided strong evidence about the role played by 'German agents' in the relations between the Kemalists and the Bolsheviks. (From Wardrop to de Robeck, 15 October, 1919; Cambridge, Churchill College Archive Centre; De Robeck Papers, DRBK 6/11.) Similarly the *Times* reported on 3 February 1920 that a number of organisations were set up in Berlin with the intention of creating a united front including the Germans, Bolsheviks, and Muslim Turkish militants. Karabekir told to Mustafa Kemal in early February 1920 that Rawlinson had a strong suspicion about 'a German–Russian–Turkish understanding'. (Karabekir, *Istiklal Harbimiz*, p. 436.)

40 On 12 August 1919 Curzon was writing to Balfour(Paris) that 'the Bolshevik armies appear to be gaining in military spirit and efficiency.'(*British Documents*, III, p. 520.)

41 Wardrop(Tiflis) to Earl Curzon, January 3, 1920, *British Documents*, III, pp. 746–7.

42 From Hardinge to Rumbold, 6 January 1920; Cambridge, Cambridge University Library; Hardinge Papers, I(1920), p. 173. (Lord Hardinge 1906–1910 and 1916–1920, permanent under-secretary of state for foreign affairs; 1920–1922, ambassador at Paris.)

43 *The Times*, 16 January 1920.

44 The Armenian Republic at Erevan was, however, given *de facto* recognition a week later. *(The Times*, 23 January 1920.)

45 Calwell, *Field-Marshall Sir Henry Wilson*, II, pp. 221–2.

46 Churchill, *The World Crisis, Part IV, The Aftermath*, p.169.

47 *Bor'ba za Ustanovlenie i Uprochenie Sovetskoi Vlasti v Dagestane 1917–1922*

gg., p. 438.

48 Statement from Narkomindel, 27 September 1919; Moscow, AVP; Fond: Reference about Turkey, Op.: 2, Por.: 1, Pap.: 2.
Letter from Chicherin to Lenin, 1 March 1920; Moscow, TsPA, Fond: 5, Op.: 2, Ed. kh.: 314.

49 Unüvar, *Istiklal Harbinde Bolseviklerle Sekiz Ay*, pp. 3–8. (This is a personal account of a Turkish officer who had been sent to Transcaucasia by Kazim Karabekir after the Erzurum Congress of 1919.)

50 Rawlinson, *Adventures in the Near East*, p. 109.

51 ibid., p. 113.

52 ibid., pp. 148–9. Rawlinson describes in his memoirs his first impression of this important fortress which stands in the shelter of mountains at a height of 7,000 feet: 'The winds there blow with terrific force, and the piercing cold defies all furs ... No tree or shrub of any sort can be found within over 50 miles, either to afford fuel when cut or shelter of any kind, and the words "dismal", "dreary", "desolate", and "damnable", suggest themselves irresistibly. (ibid., p.180)

53 *iklal Harbimiz*, p. 432; Yerasimos, *Türk–Sovyet Iliskileri*, pp. 119–21.

54 ibid., p. 156.

55 ibid., p. 221.

56 The same conclusion was reached by de Robeck, almost a month later, in 18 March 1920. (From Commander-in-Chief, Mediterranean Station, to the Secretary of Admiralty, 18 March 1920; London, PRO; FO 371/5046, E 3046/3/44/.)

57 Karabekir, *Istiklal Harbimiz*, p. 439.

58 ibid., p. 442.

59 Cited in Yerasimos, *Türk–Sovyet Iliskileri*, p. 120.

60 (Sir) Harry Luke. Political officer on the staff of Admiral de Robeck, the British high commissioner in Constantinople in 1919. Luke gives a colourful account of his experiences in Turkey and the Caucasus in his autobiography *Cities and Men*, II. (especially chapters IV–XIV)

61 Secret Telegram, No. 180, 27 May 1920; Cambridge, Churchill College Archive Centre; De Robeck Archives, DRBK 6/10.

62 Sergey Mironovich Kirov (Kostrikov), 1886–1934. After the March Revolution he became a leader of the joint Menshevik–Bolshevik organisation in Terek(northern Caucasia). (The best account on Kirov's activities in the Terek region is King's *Sergei Kirov and the Struggle for Soviet Power in the Terek Region, 1917–1918*.) He played a minor role in the Terek People's Republic of 1918, and spent most of 1919 in Astrakhan as a political commissar in the Eleventh Army. (For Kirov's activities in 1918–1919 see Kondrashev, 'S. M. Kirov, Organizator bol'shevistskogo podpol'ia v tylu u Denikina', *Voprosy Istorii*, 7(1947), pp. 3–23.) In April 1920, Kirov was appointed to the Caucasian Bureau of the Central Committee, and at the end of May became Soviet ambassador to Menshevik Georgia. He was very influential in the region in this period. *(Malen'kaia Entsiklopediia Velikaia Oktiubr'skaia Sotsialisticheskaia Revolutsiia*, p. 171.)

63 In *Bor'ba sa Ustanovlenie i Uprochenie Sovetskoi Vlasti v Dagestane 1917–1922 gg.*, pp. 316–17.

64 Karabekir, *Istiklal Harbimiz*, p. 342.

65 ibid., pp. 440–1, and 444.

66 Cited in Yerasimos, *Türk–Sovyet Iliskileri*, p.121. British forces(apart from a garrison in Batum) had been withdrawn from the Caucasus by the end of September 1919.

67 Karabekir, *Istiklal Harbimiz*, pp. 492–3.

68 Sorgun, *Bitmeyen Savas*, pp. 320–1.

69 *Times*, 8 September 1919. According to Turkish accounts the escape of Halil and Nuri was arranged by a special nationalist organisation, *Karakol. Karakol* had been set up in Constantinople after the Mudros Armistice by the members of the CUP and worked in co-ordination with the *Teskilat-i Mahsusa.* (Yerasimos, *Türk–Sovyet Iliskileri*, p.111; Criss, *Isgal Altinda Istanbul*, p. 166.)

70 From Mr Wardrop(Tiflis) to Curzon, 16 November 1919, 155471/1015/58, in *British Documents*, III, p. 649.

71 Karabekir, *Istiklal Harbimiz*, pp. 433–5; The British deciphered Mustafa Kemal's dispatch, in 'Report for week ending 13 May 1920', Appendix L; London, PRO; FO 371/E6151/262/44.

72 Immediately after the nationalist Sivas Congress, the *Narkomindel*, in a radio message, criticised the 'Anglo–French imperialism' as being the sole enemy of the eastern nationalities, and declared unconditional Soviet support for 'the emancipation and independence of the oppressed nationalities' of the East. Radio message of Narkomindel, 27 September 1919; Moscow, AVP; Fond: Reference about Turkey, Op.: 2, Por.: 2, Pap.: 2.

73 Pipes, *The Formation of the Soviet Union*, pp. 97–8, 195–9.

74 *Bor'ba za Ustanovlenie i Uprochenie Sovetskoi Vlasti v Dagestane 1917–1922gg*, p. 369.

75 His strong loyalties to both Lenin and Stalin forced him to a fatal decision a few years later in his career. For a short while in 1922–3 Ordzhonikidze was torn between Lenin and Stalin. Then he decided to stand by his countryman, Stalin. Ordzhonikidze died in 1937, suspiciously, during the second Moscow trial, just after he attempted to save his deputy Pyatakov from his countryman's terror. *(Ocherki Istorii Kommunisticheskikh Organizatsii Zakavkaz'ia*, pp. 247–52, 448–60; Lenin, *Collected Works*, XXXXV, p.788; Medvedev, *Let History Judge*, pp. 21, 193–6.)

76 From Ordzhonikidze's speech in the Council of People's Commissars, 10 July 1919, in Ordzhonikidze, *Izbrannye stat'i i rechi*, pp. 51–72.

77 Zelimkhan, a Chechen bandit of honour, was famous throughout Russia before the First World War.

78 From the Caucasian Regional Committee Meeting, Baku, May 1919; Moscow, TsPA; Fond: 5, Op.: 1, Ed. kh.: 745. Parallel to this decision within the next year separate communist parties were established for Azerbaijan (February 1920), Georgia (May 1920), and Armenia (June 1920). (Kharmandarian, *Lenin i stanovlenie Zakavkazskoi Federatsii, 1921–1923*, p. 30.)

79 Gologlu, *Erzurum Kongresi*, pp. 201–3; English translation of the National Pact in Toynbee, *The Western Question in Greece and Turkey*, pp. 209–10.

80 Two secret organisations, *Karakol* and *Teskilat-i Mahsusa*, were very active in these activities. (Zürcher, *The Unionist Factor*, p. 82.)

81 From the High Commissioner(Constantinople) to the FO, 13 February 1920; London, PRO; FO 371/5043.

82 'Note on a possible new policy in Turkey', by Mr. Ryan, British High Commission, Constantinople, 2 February 1920; and also 'Memorandum as to situation in Asia Minor', by Gordon Campbell, Intelligence Officer for the Ismid district, February 1920; London, PRO; FO 371/5042 E704/3/44.

83 A weekly summary of Intelligence Reports issued by M.I.I.c., Constantinople branch, for week ending 26 February 1920, states that 'the Turkish government has recently received from an agent at Baku a report of a conversation between himself and the President of Azerbaijani government dealing with the employment of Turkish officers as instructors to the Azerbaijani army.'(PRO, London; FO 371/5166.)

84 From Admiral Webb to Curzon, 28 February 1920; London, PRO; FO 371/5042/E 704/3/44, pp. 23–6. From de Robeck to Curzon, 13 February 1920; From W. S. Edmunds(FO) to the Secretary of the Admiralty, 10 March 1920; London, PRO; FO 371/5043.

85 From de Robeck to the Secretary of Admiralty, March 1920; London, PRO; FO 371/5046.

86 Admiral de Robeck to the Secretary of Admiralty, 18 March 1920; London, PRO; FO 371/5046.

87 From High Commissioner(Constantinople) to the FO, 11 March 1920; London, PRO; E 1357/3/44 FO 371/5043.

88 A despatch from WO to GHQ Egypt, 10 March 1920; London, PRO; FO 371/5043.

89 *The Times*, 8 March 1920; Lloyd George, *The Truth About the Peace Treaties*, II, pp. 1286–7.

90 Edmonds' Memorandum, 13 March 1920; London, PRO; FO 371/5043, E 1297/3/44/.

91 The British FO. mentioned this as a strong possibility in an earlier despatch, on 6 March 1920, to their high commissioner, which arrived in Constantinople on the next day. (Cambridge, Churchill College Archive Centre; De Robeck Papers DRBK 6/4 .)

92 Cambridge, Churchill College Archive Centre; De Robeck Papers, DRBK 6/1. For the list of those arrested and subsequently deported to Malta see London, PRO; E 2805/37/44 and E 4395/ 37/44.

93 Karabekir describes in his memoirs that he ordered Rawlinson to be taken into protective custody within the confines of his residence. Rawlinson voluntarily withdrew the Union Jack he was flying from the upper floor of his house. In his memoirs, Rawlinson seems to dispute this. (Rawlinson, *Adventures in the Near East, 1918–1922*, pp. 222–5.) Rawlinson stayed in captivity until the end of 1921, when he was exchanged with the Turkish detainees

interned by the British at Malta.

94 Cebesoy, *Milli Mücadele Hatiralari*, p. 357–8.

95 Atatürk, *A Speech Delivered by Mustafa Kemal Atatürk. 1927*, pp. 364–5. This important link between the consecutive events – the Allied occupation of Constantinople and the establishment of a provisional government in Ankara – has been missed by many historians. For instance, E. H. Carr mistakenly gives January (instead of April), 1920, as the date for the establishment of the assembly in Ankara and, therefore interprets the British occupation of Constantinople as a counter-act, a retaliation of the Allies against the establishment of an independent nationalist assembly in Ankara. (*Bolshevik Revolution*, III, pp. 249–50.)

96 According to the official Ottoman figures the population of Ankara city was 84,665 in 1914. The overall population distribution of the Ankara *vilayet* with all its surrounding *kazas* and villages was as follows: Total population: 953,817 of which 877,285 were Muslim, 20,226 were Greek, 44,507 were Armenian, 1,026 were Jews (Karpat, *Ottoman Population, 1830–1914*, pp. 172–3.)

97 *The Times*, 28 December 1923.

98 *Narkomindel*, 16 March 1920; Moscow, AVP; Fond: Near East, Op.: 3, Pap.: 2, Por.: 1.

99 From Mustafa Kemal to Karabekir, 16 March 1920, in Karabekir, *Istiklal Harbimiz*, p. 505.

100 ibid., p. 508.

101 The full content of the letter became available only recently after the opening of the Soviet Foreign Ministry documents at the *AVP* in 1992.

102 From Mustafa Kemal to the Soviet Russian government, 26 April 1920; Moscow, AVP; Fond: Reference about Turkey, Op.: 3, D.: 3, Pap.: 2.

103 Professor Stephen Blank expresses this link by saying that 'after 1920 politics in the Transcaucasus intersected with the foreign policies of the Soviet Union in the Levant and Near East'. ('Bolshevik Organisational Development in Early Soviet Transcaucasia: Autonomy v.s. Centralisation, 1918–1924', *Transcaucasia*, (ed. by R. G. Suny), p. 322.)

104 From Eliava to Narkomindel, 'Report on Turkey and Persia', September 1920; Moscow, TsPA; Lenin Files, Fond: 5, Op.: 2, D.: 92. (Shalva Zurabovich Eliava, 1883–1937. Joined the RSDLP in 1904. Active in Vologda region in 1917–1918. In 1919 member of the Revolutionary Military Councils of the Eastern and Turkestan fronts. Chairman of a special commission for Turkestan in 1919. Active in party work in Transcaucasia in 1920. (Lenin, *Collected Works*, 47, p. 572.)

105 Ivar Tenisovich Smilga (1892–1938). Old Bolshevik, active in revolutionary movement since 1907. A member of the Central Committee in 1917–1920 and then in 1925–1927. He took part in October Revolution both in Finland and in Petrograd. During the Civil War he worked at various fronts with various armies.

106 Telegramme from Lenin to Ordzhonikidze and Smilga, 17 March 1920 in Lenin, *Leninskii sbornik*, XXXIV, p. 279.

107 Mikhail Nikolaevich Tukhachevskii (1893–1937). On completion his military schooling in 1914 he was sent to the front but taken prisoner in the following year. He escaped and returned in 1917. He joined the Bolshevik Party in April 1918. Appointed commander of the Caucasian front in January 1920. He was the military chief of the whole Caucasus until April 1920 when he was sent to Poland and replaced by Levandovsky.

108 Kirov, *Stat'i, rechi, dokumenty*, I, p. 331.

109 In January 1919 the armies in the south had concentrated on two fronts: Southern Front and the Caspian–Caucasian Front. In January 1920 the Southern Front was transformed into the South-Western Front, consisting of the 12th, 13th and 14th armies; and the rest became the Caucasus Front, consisting of the 8th, 9th, 10th and 11th Armies and the Cavalry Army. (For the formation of the Red Army and the division among the fronts see Erickson, *The Soviet High Command*, chapters II and III, and Benvenuti, *The Bolsheviks and the Red Army, 1918–1922*, chapter III.)

110 At the Eight Conference of the RCP in December 1919, Anastas Mikoyan, head of the Baku Committee of the RCP, repeatedly emphasised the need for a single party organ capable of enforcing links with local Bolshevik organisations in Transcaucasia and abroad. Mikoyan had been one of the leading activists in the Baku Commune together with Stephan Shaumian. He had commanded a Bolshevik armed detachment that took part in the suppressing of the Muslim rising in Baku in March 1918. Later he had led Shaumian and the other Bolshevik Commissars through Baku as the Turks had entered it, 14–15 September 1918. (*Vos'maia Konferentsiia RKP(B) Protokoly*, pp. 38–9.)

111 In March 1921, the Caucasus Bureau of the Central Committee of the RCP(B) was split up into the Caucasus Bureau responsible for Trans–Caucasia, Daghestan and Terek region, and the South Eastern Bureau which remained in Rostov-on-Don. The Caucasus Bureau was abolished in February 1922, when the Transcaucasus Regional Committee of the Communist Party was formed. *(Bolshaia Sovetskaia Entsiklopediia*, XI, p. 124.)

112 Ordzhonikidze, *Izbrannye stat'i i rechi, 1911–1937*, p.113.

113 Nariman Kerbalai Nazaf-oglu Narimanov(1871–1925). An old Social Democrat. After March 1918 people's commissar for mining in the Baku Soviet. In 1919 and 1920 served as a Soviet official in Moscow, first as the director of Eastern Division of the Commissariat of Foreign Affairs and then as one of the heads of the Commissariat of Nationalities. Later the chairman of the Council of People's Commissars of Azerbaijan(Narimanov, *Stat'i i pis'ma*, pp. x–xiv.)

114 Polikarp Gurgenovich (Budu) Mdivani(1893–1930). An old Georgian Bolshevik, who had spent part of the World War in Persia, and after returning in 1917 had worked as a party functionary in various regions of Russia.

115 Kirov, *Stat'i, rech'i, dokumenty*, I, p. 331; Zhvaniia, *V. I. Lenin, TsK partii, i bol'sheviki Zakavkaz'ia*, p. 231.

116 Kharmandarian, *Lenin i stanovlenie Zakavkazskoi Federatsii, 1921–1923*, p. 30.

117 *Azerbaycan Kommünist Partiyasinin Tarihi*, I, p. 374.

118 Swietochowski, *Müslüman Cemaatten Ulusal Kimlige Rus Azerbaycani*, p. 236; Unüvar, *Istiklal Harbinde Bolseviklerle Sekiz Ay*, p. 18–19.

119 Kirov, *Stat'i, rech'i, dokumenty*, I, p.205; Kazim Karabekir received two reports from Dr Fuad Sabit in Baku on 26 February 1920. According to these reports the Bolsheviks had promised the Turks that they would not occupy Azerbaijan and Daghestan. Karabekir, *Istiklal Harbimiz*, pp. 459–65.

120 Karabekir, *Istiklal Harbimiz*, p. 610.

121 Iskenderov, *Iz istorii bor'by Kommunisticheskoi Partii Azerbaidzhana za pobedu sovetskoi vlasti*, p. 440; Pipes, *Formation of the Soviet Union*, pp. 226–7.

122 Yerasimos, *Türk–Sovyet Iliskileri*, p. 149; Akhmedov, *Nariman Narimanov*, p. 253; Kazemzadeh, *The Struggle for Transcaucasia (1917–1921)*, p. 284.

123 Iskenderov, *Iz istorii bor'by Kommunisticheskoi Partii Azerbaidzhana za pobedu sovetskoi vlasti*, p. 440.

124 Mir Said Sultan Galiev was a Volga Tatar, who had joined the Bolsheviks in July 1917 and remained in the party until May 1923. He was active in establishing Soviet power in the Volga region and in organising the defence of Kazan in August 1918 against the White forces. His knowledge of national and religious movements in the East won him the trust of Stalin and other leading Party and government figures. He became the highest ranking Muslim representative in the party, publisher of the official magazine of the *Narkomnats*, *Zhizn' natsional'nostei.*

125 *Zhizn'natsional'nostei*, No. 18, 1920, p. 70.

126 In a speech delivered at an All–Russia Congress of Glass and Porcelain Workers, on 29 April (the day after the Red Army entered Baku), Lenin stated that 'we know that our industry is at standstill owing to lack of fuel ... now we have an economic base that may put life into our whole industry.' (*Collected Works*, XXXI, p. 121.)

127 From *Narkomindel* to the Politburo, June 22, 1920; Moscow, TsPA; Fond: 5, Op.: 2, D.: 314.

128 From Preobrazensky (Secretary of the C.C.) to Trotsky, 10 May 1920; From Trotsky to Ordzhonikidze, 27 May 1920 in Meijer, *Trotsky Papers*, II, pp. 171–3 and 191–3.

129 The text of the armistice in Shapiro, *Soviet Treaty Series*, I, pp. 67–9.

130 Mears, *Modern Turkey*, pp. 73–4.

131 On 22 June 1920, *Narkomindel* demanded from the Politburo that 'Comrade Eliava must immediately go to Turkey' in order 'to accelerate to accomplish our projects' across the whole region of Caucasia. (Moscow, TsPA; Fond: 5, Op.: 2, D.: 314.)

132 Bekir Sami to Chicherin, 4 July 1920; Moscow, AVP; Fond: Ref. about Turkey, Op.: 3, D.: 3, Pap.:2. The arrival of Bekir Sami and his team in Moscow is also narrated in Cebesoy, *Moskova Hatiralari*, p. 61.

133 Statement from the Central Committee, July 1920; Moscow, AVP; Fond: Near East, Op.: 3, Por.: 1, Pap.: 2.

134 Politburo, July 28, 1920; Moscow, TsPA; Fond: 5, Op.: 2, D.: 314.

135 Political clauses of the Treaty of Sévres in 'Great Britain *Parliamentary*

Papers, 1920, Treaty Series No. 11, Cmd 964, pp. 16–32'. Tripartite (Sèvres) agreement on Anatolia: The British Empire, France and Italy in 'Great Britain *Parliamentary Papers*, 1920, Treaty Series No. 12, Cmd 963'.

136 Temperley, *The History of Peace Conference of Paris*, VI, p.31. This is a six-volume collection that contains all data one should need on various treaties that settled the post-war affairs.

137 Tengirsek, *Vatan Hizmetinde*, pp. 178–80; Yerasimos, *Türk–Sovyet Iliskileri*, pp. 247–8.

138 Cebesoy, *Moskova Hatiralari*, pp. 80–2.

139 Discussions among the leaders of the Turkish national movement regarding the most suitable time to start the advance over Armenia in Karabekir, *Istiklal Harbimiz*, pp., 714–18, 727–31, 764,766, 769–72, 776–7, 878, 887.

140 S. Kuznetsova's article 'Krakh turetskoi interventsii v Zakavkaz'e v 1920–1921 godakh' describes the Turkish–Armenian war as have been provoked by Western imperialist powers. *(Voprosy istorii*, no. 9, 1951, pp. 143–56.)

141 Akçura[oglu] studied at the War Academy in Constantinople, where like so many of his contemporaries he was arrested distributing Young Turk propaganda and exiled to Tripolitana in 1897. From where he escaped to Paris, whence he went back to Kazan on Volga, his native city. He became active in Russian politics, but at the same time he published a long article in the Young Turk émigré paper *Türk* (the Turk) in 1904. This article, which has been called the 'Communist Manifesto of Turkism', was titled *Uç Tarz-i Siyaset* (Three Types of Policy). For more information see Akçura[oglu], *Türkçülük. Türkçülügün Tarihi Gelisimi.*

142 In Zarevand, *United and Independent Turania*, p. 115.

143 Rawlinson, *Adventures in the Near East*, p. 229.

144 The Armenian community in Turkey had always been extremely vulnerable. Before the First World War, of the approximately two million Armenians in the Ottoman Empire, the majority were gathered in the six east Anatolian *vilayets* known collectively as 'Turkish Armenia'. In May 1915 the deportation of Armenians from the war zone in eastern Anatolia had been ordered by the Ottoman government, following the failure of the first Ottoman winter offensive in this front. It had been feared by the Turks that the Armenians, who had shown signs of sympathy for the Russians, would act as a fifth column for the Russian army. Through a combination of primitive war-time circumstances, maltreatment and ethnic and religious conflicts, it turned out to be a tragic episode: a large number of Armenians perished in the deportations. Estimates of the actual number vary between 200.000 and 1.000.000. (A survey of the controversy is given in Dyer, 'Turkish "Falsifiers" and Armenian "Deceivers": Historiography and the Armenian Massacres', *Middle Eastern Studies*, 12(1976), pp. 99–107.)

145 From the Central Committee of the Armenian Social Democratic Party to Lenin, 10 September 1920; Moscow, TsPA; Fond: 5, Op.: 2, D.: 208.

146 M. P. Pavlovich (Veltman) was Menshevik until 1917. After the

revolution he became one of few Eastern experts of the Soviet state. From 1921 he was a member of the Collegium of the Commissariat for Affairs of Nationalities. In 1922 he became the editor of the *Novyi Vostok*. *(Tarih ve Toplum*, 1980 February, p. 68.)

147 From Eliava, Skatchko, Narimanov and Pavlovich to Lenin, October 1920; Moscow, TsPA; Fond: 5, Op.: 2, D.: 208.

148 Cited in Sverdlov, *Grigorii Konstantinovich Ordhonikidze(Sergo)-Biografiia,* p. 140.

149 Boris Vasilievich Legran(1884–1936). After the October Revolution appointed deputy *Narkomvoen* (People's Commissariat for Military and Naval Affairs). In June 1919 a member of the RVS (Military Revolutionary Council) of the Tenth Army and chairman of the Military Revolutionary Tribunal of the RVSR(Military Revolutionary Council of the Republic). 1919–21, on diplomatic work in Armenia.

150 The Revolutionary Committee of Armenia (Revkom) was organised on 7 May 1920.

151 The same day, on 1 December 1920, Ordzhonikidze gave a speech in the Baku Soviet to celebrate 'Soviet Armenia'. (Ordzhonikidze, Izbr*annye stat'i i rech'i*, pp. 91–3.)

152 Veltman (Pavlovich), 'Sovetskaia rossia i anglo–frantsuzskie intrigi na vostoke', *Kommunisticheskii Internatsional*, 14(6 November 1920), p. 2947.

153 Kuznetsova, 'Krakh Turetskoi interventsii v Zakavkaz'e v 1920–1921 godakh', *Voprosy Istorii*, 9(1951), p. 147; *Ocherki Istorii Kommunisticheskikh Organizatsii Zakavkaz'e*, pp. 462–9.

154 Lenin, on 21 December 1920, at the 8th Congress of the RCP(b), states that 'The Turkish attack was planned against us. The Allies were making a pitfall for us, but fell into it themselves, because we have received Soviet Armenia. (Lenin, *Collected Works*, XXXXII, p. 246.)

155 From Erofyev(Tiflis), to Chicherin, Moscow, 24 November 1920; Moscow, AVP; Fond: Reference about Turkey, Op.: 3, Por.: 4, Pap.: 2.

156 From Bekir Sami to Chicherin, 29 November 1920; Moscow, AVP; Fond: Near East, Op.: 3, Pap.: 2, Por.: 1.

157 Zhvaniia, *V. I. Lenin, TsK Partii, i Bol'sheviki Zakavkaz'ia*, pp. 239–40; *Ocherki Istorii Kommunisticheskikh Organizatsii Zakavkaz'e*, pp. 469–76.

158 Sverdlov, *Ordzhonikidze*, pp. 143–4.

159 Lenin, *Collected Works*, XXXXII, p. 246.

160 Reviewed in Tunçay, *Türkiye'de Sol Akimlar*, p. 245–50.

161 *TBMM Zabit Ceridesi*, VIII/I/1337, p. 227.

162 From the Soviet Mission in Georgia to Chicherin, 21 November 1920; Moscow, AVP; Fond: Reference about Turkey, Op.: 3, D.: 3, Pap.: 2. From Mdivani(Ankara) to Chicherin, 20 January 1921; Moscow, AVP; Fond: Reference on Turkey, Op.: 4, Por.: 9, Pap.: 5. (This telegram describes how in Ankara the Kemalist political police arrested all prominent communists and other well-known left-wing activists. Mdivani also narrates how on one particular occasion he helped Nizamettin Bey, distinguished socialist journalist, to escape from the police and take refugee at the Soviet embassy

building in Ankara.)

163 The events leading to the murder of 15 Turkish communists in Gökay, 'Turkish Communist Party: The Fate of the Founders', *Middle Eastern Studies*, 29/2 (April 1993), pp. 220–35.

164 Telegram from the Soviet Representative in Georgia to Chicherin, 24 November 1920; Moscow, AVP; Fond: Reference about Turkey, Op.: 3, Por.: 4, Pap.: 2.

165 Telegram from Erofyev (Tiflis) to Chicherin, 24 December 1920; Moscow, AVP; Fond: Reference about Turkey, Op.: 3, Por.: 4, Pap.: 2.

166 From Chicherin to Ordzhonikidze, 18 January 1921; Moscow, AVP; Fond: Reference about Turkey, Op.: 4, D. 11, Pap.: 5.

167 Cebesoy, *Moskova Hatiralari*, pp. 123–4.

168 In a letter to the Politburo, dated by 3 January 1921, the *Narkomindel* explains that they were considering to send a representative to Turkey, who would be able to observe the military situation carefully. It was stated that some one with a sufficient military background must be chosen. (Moscow, TsPA; Fond: 5, Op.: 2, D.: 315.)

169 From Kamenev, Commander-in-Chief of All Armed Forces of the Republic, to Lenin, 17 February 1921, Meijer, *Trotsky Papers*, II, pp. 379–81.

170 In Sverdlov, *Ordzhonikidze*, p. 143.

171 From Chicherin to Krestinsky, 3 January and 8 February 1921; Moscow, TsPA; Fond: 5, Op.: 2, D.: 315.

172 Cited in Pipes, *The Formation of the Soviet Union*, p.237. David M. Lang writes that Krasin 'was given in London to understand that Baku oil ... lost much of its value without complete Russian control of the Transcaucasian pipe-line leading into Batumi over a section of Georgian territory.'(*A Modern History of Georgia*, p. 231.)

173 Maiski, 'Anglo–sovetskoe torgovoe soglashenie 1921 goda', *Voprosy istorii*, No. 5, May 1957, pp. 70–1.

174 From Krestinski(on behalf of the Central Committee) to Smilga and Ordhonikidze, 14 February 1921, in Meijer, *Trotsky Papers*, II, p. 377.

175 *Pravda*, 2 March 1921.

176 The last British detachment left Batum on 7 July 1920.

177 *TBMM Zabit Ceridesi*, IX, p. 67.

178 From Chicherin to Krestinsky, 1 March 1921; Moscow, TsPA; Fond:5, Op.: 2, D.: 315.

179 Chicherin's Note to Ali Fuat, 2 March 1921, *DVP*, III, p.556.

180 Turkish notes were briefly summarised in a letter from Chicherin to Guseynov(Baku), March, 1921; Moscow, AVP; Fond: Reference about Turkey, Op.: 4, D.: 10, Pap.: 5.

181 Cebesoy, *Moskova Hatiralari*, p. 151.

182 Given in Pipes, *The Formation of the Soviet Union*, p.238. For a useful summary on the formation of the Georgian Communist Party and establishment of Soviet power in Georgia see Sturua, 'Slavnyi put' bor'by i pobed', *Voprosy Istorii*, 5(1971), pp. 57–67.

183 Kheifets, *Sovetskaia Diplomatiia i Narody Vostoka 1921–1927*, pp. 100–4.

184 In the north Caucasus a Soviet republic of Daghestan was proclaimed in November 1920. ('Stalin's speech on the declaration of autonomous Soviet Republic of Daghestan', 13 November 1920, in *Bor'ba za Ustanovlenie Sovetskoi Vlasti v Dagestane 1917–1921 gg.*, pp. 453–4.)

Chapter 4

1 Chicherin to the Central Committee, 18 April 1921; Moscow, TsPA; Fond: 5, Op.: 2, D.: 315.

2 General accounts of British–Soviet trade negotiations include the third volume of Ullman's work on Anglo–Soviet relations, *The Anglo–Soviet Accord*, and White's *Britain and the Bolshevik Revolution* (particularly chapters 1 and 4).

3 It was only two months after the decision of the Allied Supreme Council to end their blockade of Soviet Russia. On 2 February 1920, in his report on 'the work of the All–Russia Central Executive Committee and the Council of People's Commissars delivered at the first session of the All–Russia Central Executive Committee, Seventh Convocation', Lenin read the news about 'the decision of the Allied Council adopted on January 16 to lift the blockade' and sanction trade with Soviet Russia.'(*Collected Works*, XXX, pp. 316, 560.)

4 Leonid Borisovich Krasin(1870–1926). A Marxist since the end of 1880s. In the RSDWP since the 1890s. After the Bolshevik Revolution he was at first in diplomatic work. In August 1918 he became head of the Extraordinary Commission for the Production of Articles of Military Supply for the Red Army. From 1919 onwards he was mainly engaged on diplomatic work. (For more information on Krasin see Zarnitskii and Trofimova, *Sovetskoi Strany Diplomat.*)

5 Zarnitskii and Trofimova, *Sovetskoi Strany Diplomat*, pp. 9–12.

6 Trotsky, *Stalin*, p. 1.

7 Gilbert, *Churchill*, IV, pp. 398–9.

8 ibid., p.399. Note by Curzon regarding negotiations with Krasin, 27 May, 1920, *British Documents*, XII, pp. 723–6.

9 Zarnitskii and Trofimova, *Sovetskoi Strany Diplomat*, pp. 85–98.

10 ibid., p. 95.

11 From Chicherin to the British government, 9 July 1920, *DVP*, III, pp. 21–5.

12 A detailed account of the Polish–Soviet war in Davies, *God's Playground, A History of Poland*, II, pp. 394–402.

13 Notes of a meeting held at the Villa Neubois, Spa, 9 July 1920, *British Documents*, VIII, pp. 502–6.

14 Allied Supreme Commander Marshall Foch warned that there was nothing to prevent the Bolsheviks from getting to Warsaw. *(British Documents*, VIII, p. 490.)

15 Russian translation of the letter is in *DVP*, III, pp. 54–5.

16 Chicherin's suspicion about Krasin might have played a role in this decision. A number of letters and reports of the *Narkomindel* clearly indicate

the existence in Moscow of lack of confidence in Krasin. (From Chicherin to Davidovich, 10 June 1920; from Chicherin to Krestinsky, 14 November 1920; and from Narkomindel to Politburo, 3 December 1920; Moscow, TsPA; Fond: 5, Op.: 2, Ed. kh.: 314.)

17 Lev Borisovich Kamenev(1883–1936). In April 1917 he was elected a member of the Bolshevik Party Central Committee. In 1918 Kamenev was a delegate to the Brest–Litovsk peace negotiations and also headed a Soviet mission intended for London and Paris. However, the British government expelled him and on his return he was arrested in Finland. When he returned to Petrograd in August 1918, he was elected president of the Moscow Soviet.

18 Theodore Rothstein (1871–1953), a Russian émigré, had worked as an interpreter in the British WO during the First World War. (28 October 1918; London, PRO; FO 371/3347.) 29 October – 19 November 1918 relate to a period when Rothstein was being considered for deportation. R. A. Leeper and E. H. Carr of the FO, and Basil Thomson of New Scotland Yard were involved in the decision which in the end came down against deportation. (In John Saville's introduction to Rothstein, *From Chartism to Labourism.*) Rothstein later became the Soviet ambassador in Persia in 1921.

19 On 31 August 1918 Bruce Lockhart, who had been in Russia as the official British agent, was arrested in Moscow on the charge of complicity in the counter-revolutionary conspiracies. ('Secret and Confidential Memorandum on the "Alleged Allied Conspiracy" in Russia', 5 November 1918; London, PRO; FO 371/3348, pp. 243–59.) In September in London Litvinov was arrested in retaliation and lodged in Brixton along with Bertrand Russell who was serving six months there for pamphleteering against the war, which by that time was almost won.

20 Note from the Soviet to the British government, 17 July 1920, *DVP*, III, p. 48.

21 Cited in Riddell, *Lord Riddell's Intimate Diary*, p. 221.

22 Telegram from Curzon to Chicherin, 20 July 1920, *DVP*, III, pp. 62–3.

23 ibid., pp. 61–2.

24 Fiddick, *Russia's Retreat from Poland, 1920*, p. 107. The (Second) Treaty of Riga in March 1921 established peace and a Polish–Soviet frontier considerably east of the Curzon Line.

25 Note from Curzon to Chicherin, 26 July 1920, *DVP*, III, p. 63.

26 Ullman, *Anglo–Soviet Relations*, III, pp. 272–3.

27 ibid., pp. 283–4; Andrew, *Secret Service*, p. 385.

28 Memorandum by Commander H. F. B. Maxse on the economic policy to be adopted towards Russia generally, 20 September 1920, *British Documents*, XII, p. 784.

29 In O'Connor, *The Engineer of Revolution*, p. 253.

30 'Draft Trade Agreement', 14 December 1920, *British Documents*, VIII, pp. 869–78.

31 Chicherin, *Stat'i i rech'i po voprosam mezhdunarodnom politiki*, pp. 184–5.

32 Lenin, 'Speech delivered at a meeting of activists of the Moscow

organisation of the RCP(B)', 6 December 1920, in *Collected Works*, XXXI, p. 454.

33 'Report on the Political Work of the Central Committee of the R.C.P. (B), 8 March 1921'. Lenin, *Collected Works*, XXXII, pp. 170–91.

34 'Report on Concessions delivered to the R.C.P. (B) Group at the Eight Congress of Soviets, 21 December 1920. Lenin, *Collected Works*, XXXI, pp. 463–86.

35 *The Bolshevik Revolution*, II, 195.

36 White, *The Origins of Detente*, p. 20.

37 Anglo–Soviet Trade Agreement, 16 March 1921, *DVP*, III, pp. 607–14.

38 Krasin's article 'Establishment of *de facto* relations' in *Anglo–Sovetskie Otnosheniia*, pp. 1–7.

39 Secretary's Notes of a Conference of British Ministers with the Head of the Russian Trading Delegation', 31 May 1920, at 2.45 p.m., *British Documents*, VIII, p. 288.

40 From Sir B. Thomson to Mr C. E. Heathcote-Smith, 13 June 1920; London, PRO; FO 371/5173/E 8298.

41 Zarnitskiy and Trofimova, *Sovetskoi Strany Diplomat*, p. 95.

42 Lenin, 'Report on the political work of the CC of the RCP(B)', 8 March 1921, in *Collected Works*, XXXII, p. 181.

43 From Trotsky to Chicherin, 4 June 1920, Meijer, *Trotsky Papers*, II, p. 209.

44 The English translation of the invitation was given in the Weekly Summary of Intelligence Reports Issued by S.I.S. (Constantinople Branch), for week ending 2.9.1920; London, PRO; FO 371/5177, pp. 29–30.

45 Zeki Velidi Togan(1890–1970) A Central Asian himself and a principal leader of the Turkestan National Liberation Movement.

46 Sorkin, *Pervii S'ezd Narodov Vostoka*, pp. 16–17.

47 Roy, *M. N. Roy's Memoirs*, p. 392.

48 The Baku congress is discussed in Sorkin, *Pervyi S'ezd Narodov Vostoka*; *Birinci Dogu Halklari Kurultayi - Baku 1–8 Eylül 1920. Stenoyla tutulmus tutanak*; White, 'Communism and the East: The Baku Congress, 1920', *Slavic Review*, September 1974, pp. 492–514; Tunçay, *Türkiye'de Sol Akimlar*, pp. 209–17; Aydemir, *Suyu Arayan Adam*, pp. 187–98.

49 *Izvestiia*, 21 September 1920; *Pravda*, 8 and 16 September 1920; *Kommunisticheskii Internatsional*, 14(1920), col.2941.

50 Tunçay, *Türkiye'de Sol Akimlar*, pp. 209–11.

51 Sorkin, *Pervyi S'ezd Narodov Vostoka*, p. 31.

52 Mikoyan, *Mysli i Vospominaniia o Lenine*, pp. 49–50; Tunçay, *Türkiye'de Sol Akimlar*, pp. 215–17.

53 Published in *Kommunisticheskii Internatsional*, II(20 December 1920), cols. 3141–50.

54 Mikoyan, *Memoirs of Anastas Mikoyan*, I, pp. 201–2.

55 Perhaps the only significant product of the Congress was the emergence of a special journal, *Narody Vostoka*. The first issue of this journal

came out in Russian, Turkish, Persian and Arabic in October 1920 under the editorship of M. Pavlovich.

56 Carr, *Bolshevik Revolution*, III, P. 269.

57 Weekly Summary of Intelligence Report No. 5, issued by the SIS (Constantinople Branch), for week ending 8 September 1920; London, PRO; FO 371/5177/ p. 85. Political Report on the Baku Congress, from the SIS, Source: 'A–8', 30 October 1920; London, PRO; FO 371/5178/E 13412/345/44, pp. 145–52.

58 'From Curzon to HM Representative at Constantinople, Telman, Cairo, Riga, Helsingford, Warsaw, Bucharest, Vladivostok, and Washington', 19 March 1921, *British Documents*, XII, pp. 838–41.

59 A letter from Narkomindel to the Central Committee gives a clear idea how the Bolsheviks interpreted this agreement. The letter, dated 18 April 1921, states that Enver Pasha had applied to Narkomindel for permission and support to publish two Turkish newspapers in Moscow. The aim of the papers were described as promoting anti-British feelings among the Muslims of the East, Turkey in particular. Chicherin, in his letter to the Central Committee, says that 'the agreement with Britain prohibits all propaganda activities outside the borders of Russia.' 'But', he continues, 'it does not say anything about such activities in Russia.' Therefore Chicherin affirms that there is nothing wrong in accepting Enver's proposals, on the condition that the distribution of the papers in Turkey and the other eastern countries would be done by the local communists. (From Chicherin to the Central Committee, 18 April 1921; Moscow, TsPA; Fond: 5, Op.: 2, D.: 315.)

60 *Trade Agreement between His Britannic Majesty's Government and the Government of the Russian Socialist Federal Soviet Republic*, Cmd. 1217, 1921.

61 Tengirsek, 'Milli Mücadelede Ruslarla Ilk Temaslarimiz', p.97. Ali Fuat, who also signed the treaty, maintains that treaty was signed on 18 March and upon the request of Chicherin it was agreed to publish the text of the agreement on 18 March without changing its date of signature. (Cebesoy, *Moskova Hatiralari*, p. 155.)

62 Cebesoy, *Moskova Hatiralari*, pp. 141–51.

63 From Ahmet Muhtar to Chicherin, 15 December 1920, *DVP*, III, pp.

64 From Chicherin to Ahmet Muhtar, 19 December 1920, *DVP*, III, pp. 392–6.

65 From Chicherin to Bekir Sami, 8 February 1920, *DVP*, III, pp. 512–4.

66 Bekir Sami is important as one of the very first Kemalist Foreign Office officials and diplomats. He was the first Foreign Minister of the Ankara government, and the first diplomatic negotiator sent to Moscow. ('Notes on the delegation of the Grand National Assembly of Ankara', 1 March 1921; London, PRO; FO 371/6467.)

67 The other members of the delegation were Yusuf Kemal, Minister of Economy; Dr Colonel Ibrahim Tali; Osman Bey, deputy from Trabzon; and Staff Lieutenant Colonel Seyfi. (Cebesoy, *Moskova Hatiralari*, pp. 61–2.)

68 Biyiklioglu, *Atatürk Anadolu'da*, p. 68.

69 The text of this draft treaty in ibid., pp. 80–1.

70 Cebesoy, *Moskova Hatiralari*, pp. 77–8; Tengirsek, *Vatan Hizmetinde*, pp. 178–82.

71 *TBMM Zabit Ceridesi*, III, pp. 185–90.

72 Cebesoy, *Moskova Hatiralari*, pp. 75–8.

73 Nadi, *Cerkes Ethem Kuvvetlerinin Ihaneti*, p. 11.

74 Cebesoy, *Milli Mücadele Hatiralari*, pp. 474–5.

75 Sherif Manatov was a Bashkir, the son of a mullah, and studied at the Polytechnic Institute of St. Petersburg and in Constantinople. In early 1917, Manatov was a right-wing nationalist, but later in the year he went over to the side of the Bolsheviks. Stalin rewarded him in January 1918 with one of the vice-chairmanships of the Central Commissariat for Muslim Affairs. In 1919, Manatov was sent to Ankara. His fate after 1920 is unknown. (Bennigsen, 'Marxism or Pan-Islamism', *Central Asian Survey*, VI/2(1978), p. 64.)

76 Pipes, *The Formation of the Soviet Union*, p. 158; Tunçay, *Türkiye'de Sol Akimlar*, p. 177.

77 Tunçay, *Türkiye'de Sol Akimlar*, p. 185.

78 The same agent, who provided information about the Baku Congress, gave the British SIS somewhat detail information about the First Congress of Turkish Communists. (From the SIS Constantinople Branch, 25 October 1920; London, PRO; FO 371/5178/E 13412/345/44.)

79 *Mustafa Subhi ve Yoldaslari* (*TKP* publication), pp. 65–7.

80 Demir, *Yeni Cag*, 9(September 1965), pp. 761–9.
Yakup Demir (Zeki Bastimar) was the General Secretary of the *TKP* between 1951 and 1973.

81 Cited in Ileri, *Atatürk ve Komünizm*, pp. 155–9.

82 Robert G. Wesson mistakenly states in his book *Soviet Foreign Policy in Perspective* that 'He[Mustafa Kemal] also for a time was affiliated with the Comintern through a fake Turkish Communist Party'(p. 75).

83 Cebesoy, *Milli Mücadele Hatiralari*, p. 507; Cerrahoglu, *Türkiye'de Sosyalizmin Tarihine Katki*, pp. 175–9; Harris, *Origins of Communism in Turkey*, p. 82.

84 Cebesoy, *Milli Mücadele Hatiralari*, p. 509. It is not clear from Cebesoy whether he shared the wisdom of Kemal's venture.

85 Novichev, 'Antikrest'ianskaia politika Kemalistov v 1919–1922', *Voprosy Istorii*, 9(1951), pp. 70–1.

86 According to a British intelligence report from Constantinople, during November and December 1920 there were some secret negotiations going on between the Ankara and the Greek government, and Ethem might have been chosen as Kemal's representative in these negotiations. Ethem's revolt might, therefore, have been seen as a shield to disguise these secret dealings. (Weekly Report for week ending 8 January 1921, General Staff Intelligence, Constantinople; London, PRO; FO 371/6497.) Although seems convincing, I have not come across any other account to support the existence of above mentioned secret talks.

87 *TBMM Zabit Ceridesi*, VIII/1/1337, p. 227. Kandemir, *Atatürk'ün*

Kurdurdugu TKP, pp. 134–6. This information provided by Turkish sources is confirmed by a British Intelligence report: Weekly Report No. 97. For week ending 4 December 1920. GHQ General Staff Intelligence(Constantinople); London, PRO; FO 371/6497.

88 According to one account only Mustafa Subhi's wife Semiramis survived this tragedy. Yet there is not any other available confirmation of this version. (Emre, '1920 Moskova'sinde Türk Komünistleri', *Türk Dünyasi*, 1(December 1964), p. 151.)

89 An account of this incident in Gökay, 'The Turkish Communist Party: The Fate of the Founders', *Middle Eastern Studies*, 29, 2(April 1993), pp. 220–35.

90 Summary about Turkey, Near and Middle Eastern Department, Comintern, 10 May 1922; Moscow, TsPA; Fond: 5, Op.: 3, D.: 630.

91 Tunçay, *Türkiye'de Sol Akimlar*, pp. 235–6.

92 ibid., p. 236.

93 Narkomindel Information Report summarises the details of this deadly journey, January 1921; Moscow, AVP; Fond: Reference about Turkey, No. 722.

94 Internal Party Report, RCP(B), 20 February 1921; Moscow, TsPA; Fond: 5, Op.: 2, D.:2.

95 The tragic sacrifice of the local communists to the interests of the Soviet foreign policy repeated itself on a much larger scale six years later in China. General accounts of the 'Chinese Episode' include Isaacs' *The Tragedy of the Chinese Revolution* and Trotsky's *Problems of the Chinese Revolution.*

96 Cebesoy, *Moskova Hatiralari*, pp. 102–3.

97 Moscow, AVP; Fond: 132, Ref.: Turkey, Op.: 4, D.: 11, Pap.: 5.

98 From Narkomindel to Politburo, 30 November 1920; Moscow, TsPA; Fond: 5, Op.: 2, Ed. xp.: 314.

99 Cebesoy, *Moskova Hatiralari*, pp. 144–5.

100 Among the many accounts on the amount of military aid requested by Ankara from the Soviet government are Cebesoy, *Moskova Hatiralari*, pp. 144–5, 247–8; Aralov, *Vospominaniia Sovetskogo Diplomata*, pp. 17–19; and Harris, *The Origins of Communism*, pp. 59–60.

101 Cebesoy, *Moskova Hatiralari*, p.137; Biyiklioglu, *Atatürk Anadolu'da*, p. 19.

102 Cebesoy, *Moskova Hatiralari*, p. 82.

103 Selek, *Anadolu Ihtilali*, p. 133.

104 This information (Ankara's request) is confirmed by a Narkomindel letter to Stalin on 14 September 1921; Moscow, TsPA; Fond: 5, op.: 2, d.: 315.

105 Aydemir, *Tek Adam*, II, p. 433.

106 Karal, 'Turkish Relations with Soviet Russia', pp. 270–1, 300. Ibrahim H. Karal cites the following sources: *DVP*, III, p.675; *International Affairs*, July 1960, pp. 120–2; Karabekir, *Istiklal Harbimiz*, pp. 882, 953; Selek, *Anadolu Ihtilali*, I, pp. 112–3; *Yakin Tarihimiz*, I, p.100.

107 London, PRO; FO 371/6537, E 13780/143/44.

108 This was indeed less that one-tenth of the sum demanded by the Turkish delegation. Turkish delegation demanded 150 million gold rubles which was considered by Narkomindel as a 'pure exaggeration of the Eastern mentality'. (From Narkomindel to Stalin, 10 March 1921; Moscow, TsPA; Fond: 5, Op.: 2, D.: 315.) According to Cebesoy, on the other hand, the Russians agreed to pay 10 million rubles a year for the continuation of the war with Greece. *(Moskova Hatiralari*, pp. 265–70.)

109 On 20 September 1921 Chicherin wrote to Stalin that the Turks insisted on an additional sum of 50 million gold rubles in addition to already agreed 10 million. There is no indication to whether this was accepted. (Moscow, TsPA; Fond: 5, Op.:2, D.: 315.)

110 Given in Ponomaryov, Gromyko, and Khvostok, *History of Soviet Foreign Policy, 1917–1945*, pp. 163–4.

111 The much greater amount given in a British intelligence report seems to be unfounded.

112 Professor A. Miller, prominent Soviet historian of the Eastern Affairs, writes that 'from a practical point of view, the Moscow Treaty enabled the Turks to count on the increase aid which, after the victory of the Soviet regime over the interventionists and White Guards, could now be extended in ever-growing quantities.'(Miller, *Ocherki Noveishei Istorii Turtsii*, p. 114.)

113 Lenin, 'Report on the tax in kind, delivered at a meeting of secretaries and responsible representatives of R.C.P. (B.) cells of Moscow and Moscow *gubernia*', 9 April 1921, *Collected Works*, XXXII, p. 290.

114 From Chicherin to Orakheloshvili(Tiflis), 30 March 1921; Moscow, AVP; Fond: Reference about Turkey, Op.: 4, D.: 10, Pap.: 5.

115 Tengirsek, *Vatan Hizmetinde*, pp. 293–9. An English translation of the text of Moscow Treaty in Mears, *Modern Turkey*, pp. 645–8; and in Anderson, *The Great Powers and the Near East*, pp. 175–6.

116 Spector, *The Soviet Union and the Muslim World*, pp. 85–103.

Chapter 5

1 Notes on the situation at Constantinople and in the Near East'; Cambridge, Churchill College Archive Centre; DRBK 6/13, p. 3.

2 Moscow, TsPA; Fond:5, Op.: 2, D.: 315.

3 The term was used by Curzon in a memorandum respecting intervention between Soviet Russia and Turkey, dated 7 October 1921, *British Documents*, XVII, pp. 421–33.

4 Supplementary treaty in *DVP*, IV, pp. 32–4. The Soviet government, although signed the agreement regarding the prisoners of war, was not willing to send anyone to Turkey before the conclusion of a treaty in Transcaucasia. (From Chicherin to Legran(Tiflis), 28 April 1921; Moscow, AVP; Fond: Reference about Turkey, Op.: 4, D.: 10, Pap.: 5.)

5 Aralov, *Vospominaniia Sovetskogo Diplomata*, p.32.
Semyon Ivanovich Aralov, the first head of the Soviet Military Intelligence [GRU]. Aralov, holding the top post in the GRU for about two years, had

moved down to chief of intelligence of the 12th Army in July, 1920. Since the beginning of 1921 he was a deputy of the chief of the military intelligence mainly responsible for Turkey and the Caucasus. Aralov later became the Soviet ambassador in Ankara in January 1922. Aralov established Soviet consulates at Kars, Artvin, Samsun, and Beyazit. He also placed 'agricultural experts' at Mersin, Erzurum, and Kastamonu. (ibid., pp. 67–70.)

6 At the beginning of March 1921, just two weeks before the signing of the treaty, Chicherin wrote to Krestinsky, explaining the relations with Turkey with a very negative and critical view. (From Chicherin to Krestinsky, 1 March 1921; Moscow, TsPA; Fond: 5, Op.: 2, D.: 315.)

7 Karabekir, *Istiklal Harbimiz*, pp. 891–2.

8 From Chicherin to Ali Fuat, 6 April 1921, *DVP*, IV, p. 49.

9 Karabekir, *Istiklal Harbimiz*, p.892; Kheifets, *Sovetskaia Diplomatiia i Narody Vostoka 1921–1927*, p .122.

10 Miller, A. 'The Origins of Leninist Eastern Policy', *International Affairs*, Moscow, No. 4, April 1972, p. 74.

11 Karabekir, *Istiklal Harbimiz*, p. 893.

12 From Narimanov, Head of the Azerbaijani Revkom, and Hüseynov, Azerbaijan Foreign Commissar, to the Grand National Assembly of the Turkish government, 24 April 1921, *DVP*, IV, p. 83.

13 Tengirsek, *Vatan Hizmetinde*, pp. 243–4.

14 Karabekir, *Istiklal Harbimiz*, p. 853.

15 From Chicherin to Ali Fuat, 18 May 1921, *DVP*, IV, pp. 128–9.

16 Karabekir, *Istiklal Harbimiz*, pp. 910–1.

17 *Atatürk'ün Tamim, Telgraf ve Beyannameleri*, p. 369.

18 Karabekir, *Istiklal Harbimiz*, pp. 909–10. Soviet accounts in general portray Karabekir as the one who was consistently trying to prevent further rapprochement between two sides. (Aralov, *Vospominaniia Sovetskovo Diplomata*, p. 27) The reason for that might have been related to the attempt of presenting a clear Kemal–Lenin friendship on the basis of the general principle of the solidarity of the anti-imperialist nations. Karabekir had, indeed, neither ideological nor personal sympathy to the Soviet regime. However, he should not be singled out in this position. He was a pragmatic military leader, and in no means more anti-communist than Kemal himself. (Hayit, 'Sovyet Kaynaklarinda Atatürk', *Belgelerle Türk Tarihi Dergisi*, 1 March 1985), pp. 20–5.)

19 Regarding Natsarenus's reception in Ankara see the telegram from Legran(Tiflis) to Chicherin, 7 July 1921; Moscow, AVP; Fond: Reference about Turkey, Op.: 4, D.: 10, Pap.: 5, no. 781. According to a Soviet Informburo report(Trabzon), Natsarenus arrived in Anatolia with a group of 38 men. ('From Up-to-date Review about the Situation in Anatolia', Soviet Information Bureau(Trabzon), 6 June 1921; Moscow, TsPA; Fond: 64, Op.: 1, D.: 205.)

20 From Natsarenus(Ankara) to Chicherin, 30 August 1921, in Kheifets, *Sovetskaia Diplomatiia i Narody Vostoka 1921–1927*, pp. 170–1.

21 Secret British Intelligence Report, 22 June 1922; London, PRO; FO 371/7947/E 6421. According to the Soviet sources, following the battle of Sakarya, Natsarenus asked Moscow his replacement on the ground of health conditions. This demand was accepted by the Central Committee. (From Chicherin to the Secretary of the Central Committee, 19 October 1921; Moscow, TsPA; Fond: 5, Op.: 2, D.: 315.)

22 Cebesoy, *Moskova Hatiralari*, pp. 203–9; and Aydemir, *Makedonya'dan Orta Asya'ya Enver Pasa*, p. 585.

23 From Natsarenus to Yusuf Kemal, 20 August 1921, in *DVP*, IV, pp. 287–8.

24 From Mraviyan, Armenian Foreign Commissar, to Yusuf Kemal, 24 August 1921, in *DVP*, IV, pp. 292–3.

25 Gadzhiev, *Iz Istorii Obrazovaniia i Padeniia Iugo-zapadnoi Kavkazskoi (Karskoi) Demokraticheskoi Respubliki*, p. 33.

26 *DVP*, IV, p.373; Karabekir, *Istiklal Harbimiz*, pp. 942–53, and 956–8.

27 The Central Committee of the Party especially wanted Frunze to find out detailed information about the military situation in Turkey. (See Molotov's letter to Frunze on p. 214.)

28 Mustafa Kemal, in his letter to Lenin, on 4 January 1922, describes how friendly the Turkish public received Frunze in Ankara. (Moscow, TsPA; Fond: 5, Op.: 1, D.: 1520.)

29 Treaty of Friendship and Brotherhood between Ukraine and Turkey, 2 January 1922, *DVP*, V, pp. 9–14.

30 Frunze describes his trip to Turkey(without making any reference to the strategic military situation) in his *Neizvestnoe i zabytoe*, pp. 201–39. There must be another/ confidential report mainly related to the military strategic issues, as requested from Frunze by the CC before he had left for Turkey. Unfortunately I could not find anything about it.

31 A Turkish translation of Frunze's report to the Central Committee was printed in *Frunze'nin Türkiye Anilari*, pp. 118–26.

32 *Sovyetskaia Diplomatiia i Narodni Vostoka 1921–1927*, pp. 187–8.

33 British Secretary's notes of an Allied conference held at Le Cercle Nautique, Cannes, 6 January 1922, *British Documents*, XIX, pp. 18–29.

34 The English text of the Moscow Treaty in Anderson, *The Great Powers and the Near East*, pp. 175–6; and Mears, *Modern Turkey*, pp. 645–8.

35 Aralov, *Vospomaninaniia Sovetskogo Diplomata*, p. 121.

36 From Chicherin to Krasin(London), 21 January 1922; Moscow, AVP; Fond: Near East, Op.: 7, Pap.: 8, Por.: 4.

37 Telegram from Chicherin to Krasin, 21 January 1922, *DVP*, V, p.59; an English translation of it was appended to the 'Letter from Mr Waterlow (Central European Department of the Foreign Office) to M. Krasin', 22 February 1922, *British Documents*, XIX, p. 169, f.n.2.

38 SIS to Curzon, 10 April 1922; London, PRO; FO 371/8187N3381/646/38.

39 Letter from Mr Waterlow to M. Krassin, 22 February 1922, *British Documents*, XIX, p. 169.

40 Fink, *The Genoa Conference*, p. 119.

41 On 26 July 1922, in a British FO telegram to Karakhan, Soviet Assistant Commissar for Foreign Affairs, it was stated that 'the Allied governments at Genoa invited the Russian Soviet government to co-operate with them for the purpose of establishing peace in the Near East ...' (London, PRO; FO 371/7928/E 7198/76/44, pp. 324–5.)

42 Memorandum by Mr Forbes Adam and Mr Edmonds respecting the means of bringing pressure upon Turkey, Foreign Office, 6 February 1922, *British Documents*, XVII, pp. 612–7.

43 Memorandum by Mr Ryan, 17 February 1922, *British Documents*, XVII, pp. 626–30.

44 Articles of Agreement to form the basis of a treaty with Russia, *British Documents*, XIX, pp. 199–206.

45 Draft Foreign Office reply to Karakhan, 19 July 1922; London, PRO; FO 371/7928, E 7198/76/44, pp. 324–5.

46 From Celal Bey(Ankara) to Chicherin, 10 February 1922, in Yerasimos, *Türk–Sovyet Iliskileri*, p. 459.

47 Walter Rathenau, 1867–1922; minister of reconstruction in May 1921 and minister of foreign affairs eight months later in the Weimar Republic. In June 1922 he was assassinated by anti–semitic nationalists.

48 The Rapallo Agreement threatened the new Poland and the entire East European settlement. It broke the power of the Allies' demands and represented a striking victory for the Soviets. The text of Rapallo Agreement in *Sovetsko–Germanskie Otnosheniia*, II, pp. 479–81.

49 Aralov, *Bir Sovyet Diplomati'nin Türkiye Hatiralari*, p. 133. Kazim Karabekir too followed the developments at Genoa closely and described Rapallo a very important step for the two nations, Russia and Turkey. *(Istiklal Harbimiz*, pp. 1060–1.)

50 From Karakhan to Mustafa Kemal, 27 May 1922, *DVP*, V, pp. 418–9.

51 From Karakhan (Moscow) to Aralov (Ankara), 17 June 1922, *DVP*, V, p. 454.

52 A letter from Chicherin to Stalin describes in details Enver's movements from Batum to Moscow, 11 December 1921; Moscow, TsPA; Fond: 5, Op.: 2, D.: 315.

53 The *Cheka* (All–Russian Extraordinary Commission for Combating Counter-revolution and Sabotage) was abolished on 6 February, 1922, and its functions were taken over by a newly created branch of the People's Commissariat of the Interior - *GPU* (the State Political Administration).

54 Roy, *Memoirs*, p. 408–9.

55 *GPU* was replaced by *OGPU* (United State Political Administration) on 15 January 1923.

56 Essad-Bey, *Secrets of OGPU*, p. 196. According to Mikoyan in the summer of 1922 Ordzhonikidze and Eliava were sent to the region as well. *(Memoirs of Anastas Mikoyan*, I, p. 419).

57 'Basmaci' is derived from the Turkish word *baskinci*, meaning attacker, which was first applied to bands of brigands. During Tsarist times, these

bands existed when independence was lost and Russian domination began in Turkmenistan, Bashkurdistan and the Crimea. The majority and the most influential of the Basmaci groups were founded after 1918.

58 Karabekir, *Istiklal Harbimiz*, p. 352.

59 Togan, *Bugünkü Türkili Türkistan ve Yakin Tarihi*, I, p. 435.

60 *Pravda*, 12 and 13 July 1922.

61 ibid.

62 For some months after the death of Enver the news of the incident was not accepted as true in the region. For instance, a British agent wrote to his headquarters in India that he heard that Enver had captured Bukhara and Katta-Kurgan(a provincial town between Bukhara and Samarkand). (Given by Fraser, 'Basmachi-II', *Central Asian Survey*, VI, no. 2(1987), p. 7.)

63 When he met with the Red Army petrol, he was very close to the Afghan frontier, and could easily have crossed it with his retinue had he so wanted. Whilst there is something of the Moslem precept of *sehit* (martyr) in his attitude to death on the battlefield, Enver's many letters to his wife reveal the importance which he attached to his connection by marriage with the Ottoman dynasty. On his way to Bukhara in October 1921, he had sent her a brochure in which he was described as 'Victor'[*Gazi*] and wrote: 'May the hopes expressed here be realised, so that I can show the world that I am worthy of my darling'. (Karaman, *Istiklal Mücadelesi ve Enver Pasha*, p. 99) The desire to be worthy of his 'darling' may have led to his brave but hopeless charge against the machine guns of the Red Army. 'Here', writes his biographer Aydemir, 'it was no longer a question of soldiership, but rather of the end of the road, the final effort and the quest for the expected ending'. (Aydemir, *Makedonya'dan Orta Asya'ya Enver Pasha*, III, p. 684.)

64 ibid., p. 641.

65 Togan, *Bugünkü Türkili Türkistan ve Yakin Tarihi*, I, p. 437.

66 Erer, *Enver Pasa'nin Türkistan Kurtulus Savasi*, p. 12.

67 Karabekir, *Istiklal Harbimiz*, pp. 1095–7.

68 Even before these two important appointments, Frunze pointed out in June the early indications of changing atmosphere in Ankara, saying that there was too much suspicion and a complete lack of confidence on the Turkish side. (From Frunze to Stalin, 23 June 1922; Moscow, TsPA; Fond: 5, Op.: 2, D.: 311.)

69 From M. Marx, *La perfide* (Paris, 1925) given in Yerasimos, *Türk–Sovyet Iliskileri*, p. 350.

70 Aralov's telegram was intercepted by the British, from Foreign Office Minutes, Mr Edmonds, 10 October 1922; London, PRO; FO 371/ 7902, E 10982/27/44/, pp. 128–36.

71 Aralov, *Vospomaninaniia Sovetskogo Diplomata*, p. 115.

72 *Yeni Gün*, 11 June 1922.

73 Aralov, *Vospomaninaniia Sovetskogo Diplomata*, pp. 122–3.

74 Hurewitz, *Diplomacy in the Near and Middle East*, II, pp. 98–100. A brief Soviet account of Franco–Turkish agreement in Saakyan's *Franko–Turetskie Otnosheniia i Kilikiia 1918–1923*, chapters II, III, and IV, pp. 41–179.

75 Atatürk, *Nutuk*, II, p.624; Selek, *Anadolu Ihtilali*, I, p. 667.

76 Quoted in a telegram from Rumbold(Constantinople) to Curzon, 4 January 1921; London, PRO; FO 371/6464, E 778/1/44. (Sir Horace Rumbold replaced Admiral de Robeck as high commissioner in December 1920. De Robeck still remained Mediterranean commander-in-chief.)

77 In a report from Rumbold(Constantinople), 4 January 1921; London, PRO; FO 371/6464, E 778/1/44.

78 From General Harington(Constantinople) to War Office, 14 January 1921; London, PRO; FO 371/6464. (Lieutenant-General Tim Harrington replaced Milne as British commander in September 1920.)

79 D. G. Osborne, an FO official, remarked later on 1 April 1921 that 'Kemal is brigand and patriot in much the same sense that Garibaldi was, and it is difficult not to respect and admire him.'(London, PRO; FO 371/6468/E 3565.)

80 From Rumbold(Constantinople) to Curzon, 20 January 1921; London, PRO; FO 371/6464/E 1006.

81 Cabinet meeting, 20 January 1921; London, PRO; FO 371/6464/E 1006.

82 In a 'Weekly Report for week ending 8 January 1921', GHQ General Staff Intelligence, Constantinople; London, PRO; FO 371/6497/ No. 2737 'I'. Situation Report by Richard Webb(Constantinople), No. 6., p. 15, For the period ending 25 January 1921; London, PRO; FO 371/6498.

83 'Attitude of the Turkish Nationalist Government towards Great Britain by General Staff 'Intelligence', Constantinople, December 1920–April 1921; London, PRO; FO 371/6470, pp. 185–95.

84 In a British intelligence report, 'A Secret Setting of the Great[Grand] National Assembly', (Constantinople), 25 February 1921; London, PRO; FO 371/ 6467, R/26/39.

85 Notes of an Allied Conference held at St. James's Palace, 23 February 1921, *British Documents*, XI, p. 173.

86 Montagu to Hankey, 26 February 1921; Cambridge, Trinity College Library; Montagu Papers, AS/4/8, pp. 2–3. Notes of a meeting between Lloyd George and Bekir Sami, 4 March 1921, *British Documents*, XV, p. 270.

87 Bekir Sami was described in a British FO Eastern Department report as a man of moderate political views. He is exceedingly fond of the comforts of life, on which he spends freely. He is also a confirmed gambler, so that it is not surprising that he has been at times exceedingly short of money ... It is through his constant need of money that the nationalists are able to hold him. '(19 February 1921; London, PRO; FO 371/6466, p. 90.)

88 From Foreign Office minutes, 24 February 1921; London, PRO; FO 371/6467, E 3083/1/44.

89 According to a Foreign Office minute, there were 120 Turkish prisoners held by the British at Malta, as opposed to 20 British nationals held by the Ankara government. (Foreign Office minute, 7 March 1921; London, PRO; FO 371/6499, E 3120/132/44.) The agreement in 'Vansittart-Bekir Sami Agreement', 16 March 1921; London, PRO; FO 371/6500/E 3375.

90 *The Asia Minor Expedition*, III, Appendix. 8, p. 333.

91 Toynbee, *The Western Question in Greece and Turkey*, p. 247.

92 Spyridonos, *War and Freedom*, pp. 122–7; Smith, *Ionian Vision*, pp. 198–202.

93 An account of this decision in the telegram from de Robeck(C-in-C of the Mediterranean Fleet) to British High Commissioner in Constantinople, 2 September 1921; London, PRO; FO 371/6498.

94 Memo from the C.I.G.S. on situation of British forces in and around Constantinople, 6 April 1921; London, PRO; Cab. 24/2821.

95 Rattigan(Constantinople) to Curzon, 29 May 1921; London, PRO; FO 371/6470/E 6323, p. 225. In this cyphered telegram Mr Rattigan admits Saghir's connection with British intelligence by saying that: 'Mustafa Saghir was in fact a spy. He seems to have been compromised even before he left Constantinople. His execution cannot therefore be regarded as peculiarly monstrous in itself. Its importance lies in the indication which it gives that Angora[Ankara] neither cares nor thinks it important to disguise their attitude of uncompromising hostility towards His Majesty's Government.'

96 Before proceeding to Ankara, Saghir managed to establish a contact with the secret nationalist organisation *Karakol*, which could make him able to reach the nationalist leadership in Ankara. (Criss, *Isgal Altinda Istanbul*, pp. 169–70.).

97 The Soviet government considered this incident as a clear indication of the worsening of the relations between Ankara and London. ('Up-to-date Review about the Situation in Anatolia', Soviet Information Bureau (Trabzon), 6 June 1921; Moscow, TsPA; Fond: 64, Op.: 1, D.: 205.)

98 Later in late 1922 Aralov (then the Soviet ambassador in Ankara) sent four letters to Chicherin, all dealing with the Mustafa Saghir affair. These letters describe the whole affair as a great success for the Turco–Soviet co-operation and indicate that it was the Soviet intelligence which made the Ankara government to find out the plans of the British spy. (From Aralov to Chicherin, 5 October, 25 November, 15 and 19 December 1922; Moscow, AVP; Fond: Near East, Pap.: 8, Op.: 7, Por.: 11.)

99 Curzon to the Italian Ambassador, Foreign Office, 17 June 1921, *British Documents*, XVII, p. 257.

100 Sir H. Rumbold(Constantinople) to Lord Curzon, 17 May 1921, *British Documents*, XVII, pp. 181–2.

101 'Secret Political Report on Bolshevik Activities in Constantinople', 8 June 1921; London, PRO; FO 371/6492, A/8. R/70.

102 Parliamentary question, 1 June 1921; FO 371/6470, E6323/1/44, pp. 222–3.

103 From Sir Harry Lamb's conversation with Col. Sariyannis, Smyrna, 7 June 1921, *British Documents*, XVII, pp. 228–30.

104 From Curzon to Hardinge (Paris), 14 June 1921, *British Documents*, XVII, pp. 244–8.

105 Osborne minute, 27 June 1921; London, PRO; FO 371/6519 E7247.

106 Lamb (Smyrna) to Curzon, 15 June 1921, *British Documents*, XVII, pp.

250–1.

107 In Sonyel, *Turkish Diplomacy*, p. 125.

108 Karabekir, *Istiklal Harbimiz*, pp. 948–52; Smith, *Ionian Vision*, pp. 224–5, 232–3.

109 A War Office Memorandum states that the strength of the Turkish army, at the commencement of the Greek offensive, was very much dependent upon the supplies delivered from Russia. It is estimated that 40% of the rifles, 24% of the M. G.'s, and 15% of the uniforms in the possession of the Nationalist Western Army came from the Bolsheviks. (From War Office, MI2(b), 12 December 1921; London, PRO; FO 371/6537, E 13700/143/144)

110 Enclosure 2 to a General Survey of the Military Situation at the Anatolian Front during the period from 22 July to 23 September 1921, J. H. F. McEwan, Athens, 24 December 1921; London, PRO; FO 371/7881/E322.

111 Churchill, *The World Crisis: The Aftermath*, p. 409.

112 *Turkey*, p. 100.

113 Churchill, *The World Crisis: The Aftermath*, pp. 400–2.

114 Comintern Report on Turco–Greek crisis, Berlin, 7 October 1921; Moscow, TsPA; Fond: 5, Op.: 3, D.: 628.

115 Conference of Ministers, 21 December 1921; London, PRO; FO 371/6537, p. 164.

116 This was soon accepted by the Supreme War Council. On 10 August 1921, exactly one year after the abortive peace treaty of Sèvres was signed, the neutrality of Britain, France, Italy and Japan in the Turco–Greek war was declared. (Mears, *Modern Turkey*, p. 565.)

117 Churchill, *The World Crisis: The Aftermath*, p. 393.

118 Hankey Papers, Diary, entry for 28 November 1920; Cambridge, Churchill College Archive Centre; HNKY 1/5.

119 'Nationalist Foreign Policy', from Director of Military Intelligence, 9 December 1921; London, PRO; FO 371/6537, pp. 56–61.

120 Copy of a telegram from Sir Percy Cox (Baghdad) to Secretary State for Colonies, 3 December 1921; London, PRO; FO 371/6480/E 13810.

121 Draft Conclusions of a Conference of Ministers held at 10 Downing Street, on 21 December 1921, Cabinet Minutes; London, PRO; FO 371/6481, E 14076/1/44.

122 From Cabinet Secretariat, 18 January 1922; London, PRO; FO 371/7854, E 969/5/ 44, pp. 238–41.

123 From GOC (Constantinople) to War Office, 15 and 23 December 1921; London, PRO; FO 371/6537, pp. 89–90 and 165.

124 Conversation in Paris between British, French and Italian representatives, 22–26 March 1922, *British Documents*, XVII, pp. 668–79.

125 ibid.; Rumbold (Constantinople) to Curzon, 28 March 1922, *British Documents*, XVII, p. 766.
Turkey Annual Report for 1922; London, PRO; FO 371/9176/E 10937.

126 From Greek Chargé d'Affaires to Mr. Vansittart, 1 March 1922; London, PRO; FO 371/7857, E 2588/5/44, pp. 21–2.

127 Lindley to Curzon, 26 March 1922, *British Documents*, XVII, p. 764.

128 Aralov, *Bir Sovyet Diplomatinin Türkiye Hatiralari*, p. 100.

129 Atatürk, *Speech*, pp. 545–6; Tengirsek, *Vatan Hizmetinde*, pp. 274–9; Karabekir, *Istiklal Harbimiz*, p. 1107; Biyiklioglu, *Atatürk Anadolu'da*, I, pp. 428–30; Rumbold (Constantinople) to Curzon, 5 April 1922, *British Documents*, XVII, pp. 772–4.

130 Rumbold (Constantinople) to Curzon, 23 and 25 April 1922, *British Documents*, XVII, pp. 790–1 and 793–4.

131 Lloyd George speech, 4 August 1922, *Parliamentary Debates* (House of Commons), vol. 157, col. 5.

132 Kürkçüoglu, *Türk–Ingiliz Iliskileri*, pp. 234–5.

133 Fethi's telegram to Ankara was intercepted by the British, Secret Intelligence Report, 12 July 1922; London, PRO; FO 371/7889/E 9444.

134 *Parliamentary Debates* (House of Commons), Official Report, 157, pp. 2003–4.

135 *Türk Istiklal Harbi*, II/6, book: 2, p. 277.

136 Zapantis, *Greek–Soviet Relations*, pp. 26–31; Rothschild, *The Communist Party of Bulgaria*, pp. 223–8.

137 Housepian, *Smyrna 1922: The Destruction of a City*, pp. 91–2, 209.

138 Novichev, *Turchia: Kratkaia Istoriia*, p. 161.

139 Kousoulas, *Revolution and Defeat: The Story of the Greek Communist Party*, pp. 10–1.

140 From C. H. Bentick (Counsellor of the British Embassy in Athens since 1 November 1920, acted as Chargé d'Affaires from 25 July to 17 September 1922) to Curzon, 3 September 1922; London, PRO; FO 371/7885/E 8750.

141 Curzon to Bendick (Athens), 4 September 1922, *British Documents*, XVIII, p. 5.

142 From Curzon to H. Rumbold (Constantinople), 10 September 1922, *British Documents*, XVIII, pp. 18–19; *The Communist* (English organ of the Comintern) congratulated the Turkish nationalists for their final success. (September 1922, p. 5.)

143 Given by Walder, *Chanak Affair*, p. 168.

144 London, PRO; FO 371/7885/E 8873 and FO 424/254, p. 154, no. 232.

145 British Cabinet Minutes, 7 September 1922; London, PRO; Cab. 23/31, Cabinet 48/22, pp. 1–9.

146 Gilbert, *Sir Horace Rumbold*, p. 261.

147 The chief of Imperial General Staff stated at a conference of the British cabinet ministers held on 18 September 1922 that 'The Kemalists could push that small [British] force into the sea within a fortnight. (London, PRO; FO 371/7892/E 9770/ G.)

148 From War Office to General Harington(Constantinople), 11 September 1922; London, PRO; FO 371/7872. From H. Rumbold (Constantinople) to Curzon, 14 September 1922; London, PRO; FO 424/254, p. 210. From General Harington (Constantinople) to War Office, 14 September 1922;

London, PRO; FO 371/7889/E 9492.

149 Price, *Extra-Special Correspondent*, p. 126.

150 ibid., p. 127.

15 From the minutes and conclusions of the Conference of British Cabinet Ministers, 18 September 1922; London, PRO; FO 371/7892/E 9770/G.

152 In Gilbert, *Churchill*, p. 450.

153 Conversation between the Prime Minister and M. Diamandy, 1 October 1922; London, PRO; FO 424/255, pp. 43–4.

154 Note from Karakhan, Vice-Commissar for Foreign Affairs, to Curzon, 12 September 1922, *DVP*, V, pp. 574–7.

155 From GOC Allied Forces in Constantinople to War Office, 18 September 1922; London, PRO; FO 371/7891/E 9677.

156 A Report by the British Secret Intelligence Service, 22 September 1922; London, PRO; FO 371/7893.

157 A Report by the British Secret Intelligence Service, 25 September 1922; London, PRO; FO 371/7896.

158 Note from Karakhan to the Foreign Ministers of Great Britain, France, Italy, Yugoslavia, Bulgaria, Rumania, and Greece, and the Prime Minister of Egypt, 24 September 1922, *DVP*, V, pp. 593–5.

159 Aralov, *Bir Sovyet Diplomatinin Türkiye Hatiralari*, p. 149.

160 A Report by the British Secret Intelligence Service, 22 September 1922; London, PRO; FO 371/7893.

161 For the background of the negotiations with Urquhart see:
From Lenin to Krzhizhanovsky (head of the State Planning Commission), 22 August 1921; From Lenin to Chicherin, 16 October 1921; Lenin's markings on a memo from and letter to Chicherin, 22 October 1921; From Lenin to Zinoviev, 18 September 1922, all in *Collected Works*, XXXXV, pp. 261, 339–40, 354–5, 567–8.

162 Urquhart had been born in Turkey and educated in Constantinople, Glasgow and Edinburgh. Between 1896 and 1906 he worked for his father's oil business in Baku, where he also became British Vice-Consul. In 1906 he obtained an option on the Kystym estate in the Ural Mountains near Ekaterinburg. Before the Bolshevik Revolution he had controlled a wide range of mineral resources in Russia through his company 'Russo–Asiatic Consolidated Ltd.' In 1918, the British Department of Overseas Trade authorised Urquhart to establish a company, acting as an agent of the British government, to sell goods in areas under the control of Admiral Kolchak's White forces in Siberia. In this role, he had an amicable relationship with Kolchak and the intervening Allied authorities. (MacDonell, *And Nothing Long*, p. 152; Küçük, *Türkiye Üzerine Tezler*, pp. 55–6; 'Addenda to the Terms of the Agreement with L. Urquhart', 25 October 1922, in Lenin, *Collected Works*, XXXXII, p. 424).

163 Roland McNeill on 11 April 1923 said in the House of Commons that Mr Hodgson 'has no regular diplomatic status; his exceptional position is governed by the Trade Agreement of 16 March 1921.'(*Parliamentary Debates*,

House of Commons, vol. 162, p. 1181).

164 Hodgson (Moscow) to Curzon, 20 September 1922, *British Documents*, XVIII, pp. 36–7.

165 S.I.S., 22 September 1922; London, PRO; FO 371/7893/ No. 887.

166 'A Report by the British SIS, The Near Eastern Crisis: The Question of Russian Intervention', 25 September 1922; London, PRO; FO 371/7896/ No. 892.

167 From Rumbold (Constantinople) to Curzon, 25 September, 1922; London, PRO; FO 424/254, p. 332, No. 610.

168 *§TBMM Gizli Celse Zabitlari*, III, p. 813.

169 General Harington (Constantinople), to War Office, 27 September 1922; London, PRO; FO 371/7896/ E 10195.

170 Rumbold (Constantinople) to Curzon, 27 September 1922, *British Documents*, XVIII, p. 104.

171 Minutes of a Conference of British Ministers held at Lord Curzon's house, 29 September 1922; London, PRO; FO 371/7898/ E 10399.

172 From Rumbold (Constantinople) to Curzon, 2 October 1922; London, PRO; FO 371/ 7897/ E 10366. Atatürk, *Speech*, pp. 568–9; Cebesoy, *Siyasi Hatiralar*, pp. 72 and 75–9.

173 From the meeting between Poincairé, Curzon and Galli (Italy's Paris representative), 7 October 1922, *British Documents*, XVIII, pp. 165–71.

174 G.H.Q. (Constantinople) to War Office, 29 September 1922; London, PRO; FO 371/7897.

175 Conversation between the British Prime Minister and Romanian representative M. Diamandy, 1 October 1922; London, PRO; FO 424/255, pp. 43–4. No. 55.

176 The original armistice text in French, London, PRO; FO 371/ 7905/ E 11487. In Turkish in Türkgeldi, *Mondros ve Mudanya Mütarekeleri Tarihi*, pp. 178–81. In English in Mears, *Modern Turkey*, pp. 658–9. As the peace talks opened at Mudania, a coup d'état in Athens led by Venizelist officers toppled King Constantine. He was replaced by George II as the Venizelist promised to punish those responsible for the debacle in Anatolia.

Chapter 6

1 Memo by H. Nicolson, Foreign Office, 3 October 1922; London, PRO; FO 371/ 7898, pp. 220–35.

2 Cited in Conte, *Christian Rakovski (1873-1941)*, p. 218.

3 'Angora and the East', *The Contemporary Review*, June 1923, p. 688.

4 Biyiklioglu, *Atatürk Anadolu'da*, pp. 450–4.

5 Curzon to Lord Hardinge(Paris), 12 October 1922, *British Documents*, XVIII, pp. 188–91.

6 Lausanne had been the seat of another peace conference between the Ottoman Empire and Italy in 1912. This conference resulted in the Treaty of Lausanne(Ouchy), 18 October 1912. (Ahmad, *The Young Turks*, pp. 112–13.)

7 British Secretary's Notes of a Conference between France, Britain and Italy, 20 September 1922, *British Documents*, XVIII, p. 61.

8 From Karakhan to the governments of Great Britain, France, Italy, Yugoslavia, Bulgaria, Romania and Greece, and Foreign Minister of Egypt, 24 September 1922, *DVP*, V, pp. 593–95. Also see the 'Record by Mr Gregory (Head of the Northern Department at the FO) of a conversation with Berzin (Assistant Official Agent of the Soviet government in Great Britain), 26 September 1922, *British Documents*, XVIII, pp. 102–3.

9 *Pravda*, 14 September 1922.

10 *Pravda*, 20 September 1922.

11 *Pravda*, 25 September 1922.

12 *Pravda*, 8 October 1922.

13 From Karakhan to Curzon, 12 September 1922, *DVP*, V, pp. 574–7.

14 Rumbold (Constantinople) to Curzon, 5 October 1922, *British Documents*, XVIII, p. 141.

15 *Türk Tarih*, IV, p. 126.

16 Telegram from Chicherin to the British and Italian governments, 19 October 1922, *DVP*, V, pp. 621–3.

17 Berzin to Karakhan, 27 September 1922, *DVP*, V, p. 598.

18 From Aralov to Chicherin, 5 October 1922; Moscow, AVP; Fond: Near East, Op.: 7, Pap.: 8, Por.: 11.

19 Curzon to Peters(Moscow), 10 November 1922, *British Documents*, XVIII, pp. 248–9. (Mr Peters – assistant agent of the British Commercial Mission to Russia from July 1921.)

20 Mr Peters(Moscow) to Curzon, 27 October 1922, *British Documents*, XVIII, pp. 211–2.

21 Interview given to M. Farbman, *Observer* and *Manchester Guardian* correspondent in Moscow, 27 October 1922, Lenin, *Collected Works*, XXXIII, pp. 387–8. Krasin's prestige received a severe setback as a result of non-ratification of the agreement. At the Twelfth Party Congress in April 1923, Krasin was sharply criticised for having been the sole proponent for ratification of the Urquhart Concession. Urquhart, on the other hand, was keen on doing business in the region. After the failure of his agreement with the Bolsheviks he turned to Turkey, his birth place. In January 1923 while the negotiations were going on at Lausanne, Urquhart set up a company in Ankara with the extensive rights in the fields of mining and rail-roads. (Küçük, *Türkiye Üzerine Tezler*, pp. 55–61.)

22 Martin, 'The Urquhart Concession and Anglo–Soviet Relations', *Jahrbücher für Geschichte Osteuropas*, No. 20, p. 566.

23 English translation of minutes of the several meetings at the Fourth Congress of the Comintern in SIS Report, 13 December 1922; London, PRO; FO 371/ 8180.

24 *Bulletin of the IVth Congress of the Communist International*, XVI(25 November 1922), Moscow, pp. 27–31.

25 *Bulletin of the IVth Congress of the Communist International*, XVII, pp. 27–31.

26 ibid.

27 Communist International Information Bulletin, by the Secretary of the Eastern Bureau, 21 October 1922; Moscow, TsPA; Fond: 5, Op.: 3, D.: 629.

28 From Zinoviev to Lenin, Trotsky, Radek and Bukharin, 14 November 1922; Moscow, TsPA; Fond: 5, Op.: 3, D.: 141.

29 Aralov, *Bir Sovyet Diplomatinin Türkiye Hatiralari*, p. 38.

30 Bagirov, *Iz istorii sovetsko – turetskikh otnoshenii v 1920–1922 godakh*, p. 94.

31 From Aralov(Ankara) to Chicherin, 5 October 1922; Moscow, AVP; Fond: Near East, Op.: 7, Pap.: 8, Por.: 11. From Rosenberg(Ankara) to Narkomindel, 15 October 1922; Moscow, AVP; Fond: Near East, Op.: 7, Pap.: 8, Por.: 7. Comintern (Eastern Bureau) Information Bulletin, 21 October 1922; Moscow, TsPA; Fond: 5, Op.: 3, D.:629.

32 From Aralov(Ankara) to Chicherin(Moscow), 4 November 1922; Moscow, AVP; Fond: Near East, Op.: 7, Pap.: 8, Por.: 11.

33 From Chicherin to Aralov(Ankara), 3 November 1922, Moscow, AVP; Fond: Near East, Op.: 7, Por.: 4, Pap.: 8.
From 'General Observations on the Lausanne Talks' by Chicherin, 20 February 1923; Moscow, TsPA; Fond: 5, Op.: 1, D.: 1990.

34 Aralov, S. I. *Bir Sovyet Diplomatinin Türkiye Hatiralari*, p. 173–4.

35 'Türk Milletine Sulh, Avrupa Emperyalizmine Karsi Muharebe' [Peace to the Turkish People, Struggle Against the European Imperialism], Third (Communist) International, 11 October 1922, *Ziya*, 47, p. 1.

36 Sadrettin Celal, Fourth Congress of the Comintern, 20th session(23 November 1922), *Bulletin of the IVth Congress of the Communist International*, XX, pp. 11–15.

37 273 of the 335 Conservative MPs met at the Carlton Club in London to discuss whether or not to remain in the coalition. 185 of those present voted to withdraw their support from Lloyd George. Only 88 wished the coalition to continue. (Gilbert, *Churchill*, p. 453; Morgan, *Consensus and Disunity*, pp. 280–356.; see also Appendix I: 'The Carlton Club Meeting' in Kinnear, *The Fall of Lloyd George*, pp. 221–42.)

38 Busch, *Mudros to Lausanne*, p. 360. A detailed account of the events which led to the fall of Lloyd George in Kinnear, *The Fall of Lloyd George*.

39 Memorandum by the General Staff on the Proposed New Treaty between the Allies and Turkey, War Office, 19 October 1922, *British Documents*, XVIII, pp. 984–9.

40 Nicolson, *George Curzon: The Last Phase, 1919–1925*, p. 282.

41 For details of how Curzon assumed the presidency of the all important First Commission on Territorial and Military Affairs see Dakin, D. 'Lord Curzon's policy towards Greece', *Essays in Memory of Basil Laurdas*, pp. 552–3.

42 A detailed account of this phase in Vat, *The Ship That Changed The World*.

43 Memorandum by Mr H. G. Nicolson respecting the 'Freedom of the Straits', Foreign Office, 15 November 1922, *British Documents*, XVIII, pp.

974–83.

44 Lenin, 'Interview given to Michael Farbman, *Observer* and *Manchester Guardian* correspondent', 27 October 1922, *Collected Works*, XXXIII, pp. 383–9.

45 In addition to Georgi Chicherin the Russian delegation included Chretien Rakovski, commissar for foreign affairs for the Ukraine, Polycarpe Mdivani, commissar for foreign affairs for Georgia, and Vatslav Vorovsky, Soviet official in Rome.

46 *Izvestiia*, 14 December 1922.

47 Bilsel, 'International Law in Turkey', *American Journal of International Law*, XXXVIII, 4(October 1944), pp. 550–2.

48 Cited in Conte, *Christian Rakovski*, p. 225.

49 Curzon(Lausanne) to E. Crowe(London), 6 December 1922, *British Documents*, XVIII, pp. 370–1.

50 For the Turkish position see telegram from Ismet(Lausanne) to GNA (Ankara), 5 December 1922, *Lozan Telgraflari, I*, No. 82, pp. 167–70.

51 Apart from Ismet Pasha the Turkish delegation included Dr Riza Nur, the minister of public health and deputy from Sinop; Hasan Bey, ex-minister of economy and deputy from Trabzon; Zekai Bey, deputy from Adana; Celal Bey, deputy from Saruhan; Zülfi Bey, deputy from Diyarbakir; and Veli Bey, deputy from Burdur. (Soysal, *Dis Politika ve Parlemento*, p. 93; Grew, *Turbulent Era*, I, p. 481.)

52 Some writers, notably Vere–Hodge, have stated that the Turkish nationalists 'were unanimous in appointing him [Ismet] to the important role of chief delegate at Lausanne'(*Turkish Foreign Policy*, p. 38). According to a number of eye-witness accounts, however, Ismet, only 34 years old, was appointed leader of the Turkish delegation to the surprise of everyone including himself. 'This decision was taken by Mustafa Kemal himself', partly because Ismet was his most loyal and dependable supporter, but also because then the prime minister(Rauf Bey) was known as an Anglophile, while the minister of foreign affairs(Yusuf Kemal) was too pro–Soviet. ('Rauf Orbay'in Hatiralari', *Yakin Tarihimiz*, IV, p. 19.)

53 *Pravda*, 28 November 1922.

54 From Ismet(Lausanne) to Ankara, 2 December 1922, *Lozan Telgraflari*, I, No. 68, 159.

55 Telegram from the Soviet delegation in Lausanne to Karakhan, 23 December 1922, *DVP*, VI, pp. 111–2.

56 Dontas, *Greece and Turkey*, p. 164.

57 From Ismet Pasha to the Office of the Prime Minister(Ankara), 2 February 1923, *Lozan Telgraflari*, I, No. 481, p. 482. The Lausanne Straits Convention lasted only 13 years. On 26 July 1936, it was replaced by the Montreux Conference which today is still in effect.

58 The text of the Convention in Howard, *The Problem of the Turkish Straits*, pp. 21–4; and Hurewitz, *Diplomacy in the Near and Middle East*, II, pp. 124–7.

59 Dontas, *Greece and Turkey*, p. 165; Yerasimos, *Azgelismislik Sürecinde*

Türkiye, III, p. 1233.

60 Conte, *Christian Rakovski*, p. 227.

61 Rumbold(Lausanne) to Curzon, 20 July 1923, *British Documents*, XVIII, p. 971.

62 *Istoriia Vneshnei Politiki SSSR*, pp. 175–6.

63 Maurice Alexandre Conradi was a *Russlandsschweizer*, a Swiss who had lived in Russia. He had born in 1896 in St. Petersburg where his grandfather had founded a chocolate factory. After the Bolshevik Revolution the family had lost all its fortune. Conradi, who had joined the Russian army in 1914, had joined the Volunteer Army following the Bolsheviks' seizure of power. After the collapse of the White armies he had left Russia and stayed in Turkey with thousands of other Russian refugees. After spending seven months at the refugee camps in Gallipoli Conradi went to Switzerland in the second half of 1921. (Senn, *Assassination in Switzerland, The Murder of Vatslav Vorovsky*, pp. 35–52.)

64 *Izvestiia*, 12 May 1923; From Piyashev's 'Forward' to *V. V. Vorovsky*, p. 26; Aralov, S. I. *Bir Sovyet Diplomatinin Türkiye Hatiralari*, p. 219; Carr, *A History of Soviet Russia, Vol. IV, The Interregnum, 1923–1924*, pp. 169–74.

65 Note from Chicherin to the President of the Lausanne Conference on the Assassination of Vorovsky, 16 May 1923, *Izvestia*, 19 May 1923. English translation in Degras, *Soviet Documents on Foreign Policy*, I, pp. 392–4.

66 In Moscow on 10 and 12 May large anti–British demonstrations were organised with the object of protesting against the murder of Vorovsky. *(Pravda*, 19 May 1923.) The report and the photographs of these demonstrations were sent to the FO by Hodgson on 15 May. (Two telegrams from Hodgson to Curzon, 15 and 17 May 1923, *British Documents*, XXV, pp. 119–20, and 128–30.)

67 Carr, *The Interregnum 1923–1924*, pp. 165–73; White, *Britain and the Bolshevik Revolution*, pp. 150–71.

68 On 3 October 1920, eight months after the execution of Davison, Chicherin wrote to the Central Committee, informing that a special committee was appointed to investigate the execution of Davison in Petrograd on 17 January the same year. (Moscow, TsPA; Fond: 5, Op.: 2, Ed. xp.: 314.)

69 Harding, 'The Case of Mrs Stan Harding' and 'The Moscow Trial', *The Nineteenth Century and After*, July–August 1922, pp. 1–16, 280–7.

70 *Parliamentary Debates*, 163, col.21, 424; and 166, col.924.

71 *Izvestiia*, 6 May 1922.

72 In 1921, Patriarch Tikhon (Belavin, Vasily Ivanovich, 1865–1925) had organised a church campaign for the benefit of hungry people in the Volga area. At this time the Soviet government had not made any objection to Tikhon. Tikhon had even been allowed to issue an appeal for help on radio.

73 'Correspondence between His Majesty's Government and the Soviet Government respecting the relations between the two governments', *Command Paper*, Cmd. 1869, pp. 3–4. Budkiewizc was shot on 31 March 1923. (From W. Max Muller in Warsaw to Curzon, 11 April 1923, *British*

Documents, XXV, p. 69.)

74 On 30 May 1923, in the House of Commons, Ronald McNeill, the under-secretary of state for foreign affairs, outlined the debate on the territorial waters and stated that 'a belt wider than three miles has been claimed for fishery purposes at various times by Spain, Norway, Sweden, Denmark, Russia and Uruguay'. *(Parliamentary Debates*, 164, p. 1261.)

75 Roland McNeill, 7 March 1923, *Parliamentary Debates*, 161, p. 462.

76 From Hodgson(Moscow) to Chicherin, 28 April 1923, *British Documents*, XXV, p. 87.

77 Cited in Conte, *Christian Rakovski*, pp. 256–7.

78 *Pravda*, 10 and 12 May 1923.

79 The British memorandum was discussed in the cabinet on 2 May, and sent to Hodgson in Moscow the next day. (From Curzon to Hodgson, 2 May 1923, *British Documents*, XXV, p. 88.) The memorandum was delivered to Litvinov, Assistant People's Commissar for Foreign Affairs, on 8 May. (From Hodgson to Curzon, 8 May 1923, *British Documents*, XXV, p. 100.)

80 It was the reports from SIS and the Indian intelligence bureaux, and intercepted telegrams exchanged between Moscow and the Soviet representatives in Persia and Afghanistan which made Curzon sure about the activities of the Soviet agents distributing large amounts of money and propaganda to anti-British national movements. (Andrew, 'The British Secret Service and Anglo–Soviet Relations in the 1920s. Part I: From the Trade Negotiations to the Zinoviev Letter', *The Historical Journal*, XX, 3(1977), p. 692.)

81 *Parliamentary Papers*, Command paper, Cmd. 1890 (1923), pp. 6–13. A Russian translation of 'Curzon's Ultimatum' was printed in *Anglo–Sovetskie Otnosheniia*, pp. 32–9.

82 From C. Barclay(Stockholm) to Curzon, 11 May 1923, *British Documents*, XXV, p. 108.

83 *Anglo–Sovetskie Otnosheniia*, p. 32–47; *Istoriia Vneshnei Politiki SSSR*, pp. 176–8.

84 When Curzon died in 1925, Radek gave a stinging portrait of his policy: 'The fear of Russian tsarist imperialism constituted for this representative of British imperialism the central point of his policy. To prevent Russia from penetrating into Asia was the dominating idea of Curzon ... He hated Russia, even quite independent of the class which was ruling it. He hated the Russian people in general because of that role which it was called upon to play in the awakening of Asia, this selected object of English exploitation'. (Radek, 'Lord Curzon and the Soviet Union', *Labour Monthly*, Vol. 7, No. 5, pp. 270–4.

85 Note from Litvinov to the British government, 11 May 1923, *DVP*, VI, pp. 327–30.

86 Note from Krasin to the British government, 23 May 1923, *DVP*, VI, pp. 325–7.

87 'British Memorandum', 29 May 1923, *DVP*, VI, pp. 327–30.

88 Soviet Memorandum to the British government, 4 June 1923, *DVP*, VI, pp. 334–8.

89 English translation of it in *Parliamentary Papers*, Command paper, Cmd. 1890 (1923), p. 12.

90 ibid., pp. 13–14. A Russian summary of the British reply of 13 June in *DVP*, VI, pp. 338–9.

91 Chicherin's Note to Curzon, 18 June 1923, *DVP*, VI, pp. 353–4.

92 Soysal, 'Seventy Years of Turkish–Arab Relations and an Analysis of Turkish–Iraqi Relations (1920–1990)', *Studies on Turkish–Arab Relations*, 6 (1991), pp. 27–9.

93 For the text of the treaty see *Parliamentary Papers*, 1930, Treaty Series, No. 7, Command Papers, Cmd. 3488.

94 A. Emin [Yalman], *Vatan*, 21 July and 2 August, 1923, in *Tarih ve Toplum*, 124(April 1994), pp. 54–7; Smith, *Ionian Vision*, p. 335. The Comintern and the Communist Party of Bulgaria (as the leading CP in the Balkans) strongly protested this population transfer. (Rothschild, *The Communist Party of Bulgaria*, pp. 234–5.)

95 Smith, *Ionian Vision*, p. 334.

96 For the history of the capitulations see Sousa, *Capitulatory Regime of Turkey*.

97 Hurewitz, *Diplomacy in the Near and Middle East*, II, pp. 119–27.

98 Mears, *Modern Turkey*, p. 664.

99 The decision on reparations was explained in a letter (8 August 1923) written by S. D. Waley of the Treasury to L. Oliphant of the FO in following words: 'a bird in the hand is worth any number of birds in the bush.' (Cambridge, UL; Baldwin Papers, 111, F1/ Series A, 9. Turkey, p. 226.)

Conclusion

1 Churchill, *The World Crisis, Part IV, The Aftermath*, p. 438; Mosley, *Curzon, End of an Epoch*, p. 251; Sacher, p. 442.

2 Toynbee, 'The New Status of Turkey', *Contemporary Review*, London 1923, p. 288.

3 Memorandum by H. G. Nicolson respecting the Freedom of the Straits, F.O., 15 November 1922, *British Documents*, XVIII, pp. 974–83.

4 Nicolson, *George Curzon: The Last Phase, 1919–1925*, p. 307.

5 Toynbee and Kirkwood, *Turkey*, pp. 67–8.

6 Memorandum by Lindsay (British Foreign Office), 22 February 1926; London, PRO; Cabinet Papers, Cab. 27/312, I.P(26). 4.

Bibliography

Archival Sources (Unpublished)

Great Britain

a) Cambridge University Library

Baldwin Papers: 111 (Foreign Affairs, Turkey)
Hardinge Papers: 42–5 (Personal correspondence, 1920–1922)

b) Churchill Archives Centre, Churchill College, Cambridge

De Robeck Papers: DRBK 6/ 1, 2, 3, 4, 8, 9, 10, 11, 13, 14, 15, 16, 18, 19,20, 21, 22, 23, 24, 25, 26, 28, 29, 30. (Private correspondence, telegrams, papers on foreign policy, intelligence reports, 1918–1922)
The Hankey Papers: HNKY 1/ 5, 6. (Diaries and loose diary pages, 1918–1922)
Archives of General Lord Rawlinson (Supreme War Council, Allied military representative in Russia, 1918–1919): RWLN 1\2 (1914–1924, various documents)
Archives of Lord Vansittart of Denham (First Secretary at Paris Conference, and Secretary to Foreign Secretary in 1920): VNST II/ 2/ 2, 3; 6/ 6 (1919–1921)

c) Public Record Office, London.

Cabinet Office Papers: Cab. 21, 23, 24, 29, 31 (1918–1923)

Foreign Office, Confidential Print, Eastern Affairs: FO 406/40–57 (1918–1922)
Foreign Office, Confidential Print, Russia and Soviet Union: FO 418/53–9 (1919–1923)
Foreign Office, Confidential Print, Turkey: FO 424/254–9 (1922–1923)
Foreign Office, Political Correspondence: FO 371/ 3290, 3291, 3292, 3293, 3294, 3295, 3345, 3346, 3347, 3348, 3349, 4975, 4976, 4977, 5032, 5042, 5043, 5044, 5046, 5047, 5054, 5057, 5058, 5089, 5138, 5166, 5167, 5168,

5170, 5171, 5172, 5177, 5178, 5179, 5214, 5260, 5270, 5273, 5287, 5288, 5289, 5290, 5291, 5292, 5293, 6463, 6464, 6465, 6466, 6467, 6468, 6469, 6470, 6471, 6472, 6475, 6480, 6481, 6491, 6492, 6497, 6498, 6499, 6503, 6530, 6535, 6536, 6537, 6574, 7725, 7726, 7727, 7728, 7853, 7854, 7857, 7859, 7860, 7861, 7862, 7870, 7875, 7880, 7885, 7891, 7896, 7897, 7898, 7899, 7900, 7901, 7902, 7910, 7911, 7919, 7920, 7921, 7924, 7928, 7930, 7935, 7940, 7944, 7946, 7951, 7955, 7959, 7961, 7969, 8152, 8165, 8183, 8180, 8215, 8216, 8842, 9058, 9059, 9100, 9101, 9130, 9175, 9176, 9275, 9332, 9340, 9374 (1918–1923)

Foreign Office, Private Collections: FO 800/430(Toynbee) 199, 205 (Balfour).

War Office: WO 78 (Maps and Plans)
War Office: WO 95/4960 (War Diaries)
War Office: WO 106/ 38, 39, 183 (Russia)
War Office: WO 1190 (Misc. Papers and Telegrams)
War Office: WO 1206 (Record of Naval Events in the Caspian)
War Office: WO 1493 (Turkey and Greece, Situation);
War Office: WO 157/765–7 (South Russia Intelligence Summaries).

d) Trinity College Library, Cambridge.

Papers of Edwin S. Montagu: AS 1/3. (Correspondence with Churchill, 1921)
AS 3/ 1, 10 (Various correspondence, 1919–1922)
AS 4/ 4, 5, 6, 7, 8 (Correspondence on Turkey, 1918–1921)
AS 5/9 (Correspondence with Agha Khan, 1921)

Russia

a) The Arkhiv Vneshnei Politiki. (AVP SSSR) Moscow.

Foreign Affairs Correspondence: Fond: Reference on Turkey
Opis: 1/ Delo: 2, 3 (1918);
Opis: 2/ Delo: 1 (1919);
Opis: 3/ Delo: 2, 3, 4(1920);
Opis: 4/ Delo: 1, 2, 3, 4, 5, 6, 7, 8, 9, 10, 11, 12 (1921)

Foreign Affairs Correspondence: Fond: Near East
Opis: 3/ Delo: 1 (1920)
Opis: 7/ Delo: 1, 2, 4, 5, 6, 7, 11 (1922)

Calendar of Events in Turkey (1918–1925)

b) Tsentral'ny Partiiny Arkhiv Instituta Marksizma–Leninizma pri Tsentral'nom Komitete KPSS (TsPA IML) Moscow.

Lenin Papers
Fond: 2/ Opis: 3/ Delo: 682, 689 (1920)
Fond: 5/ Opis: 1/ Delo: 821, 1520, 1990; Opis: 2/ Delo: 92, 108, 208, 210, 311, 314, 315; Opis: 3/ Delo: 141, 213, 222, 357, 367, 364, 365,503, 608, 628, 629, 630 (1920–1923)

Papers of Kavburo
Fond: 64/ Opis: 1/ Delo: 80, 112, 135, 136, 138, 205, 208, 210 (1920–22); Fond: 69/ Opis: 1/ Delo: 207 (1921)

Summary Reports by the Orgburo of the Communist Party of Turkey, Nos. I–V (1921)

Official Publications

American Papers

Department of State, *Foreign Relations of the United States, 1918, Russia*, Washington: US Government Printing Office, Vol. I (1931) II (1932) III (1932).

Department of State, *Foreign Relations of the United States, 1919, Russia*, Washington: US Government Printing Office, 1937.

Department of State, *Foreign Relations of the United States: The Paris Peace Conference, 1919*, Washington: US Government Printing Office, Vol. III (1943).

Department of State, *The Problem of the Turkish Straits*, Publication no. 2752 (compiled by) H. N. Howard, Washington: US Government Printing Office, 1947.

Harbord, Major-General J. G. *A Report of the American Military Mission to Armenia*, US Congress. 66th Cong., 2nd Session. Senate. Document No. 266, (1920).

United States Military Intelligence (Weekly Summaries) New York and London: Garland Publishing, INC., Vols. I–XII (1978) Vols. XIII–XXII (1979).

British Papers

British and Foreign State Papers, 1917–1918, Vol. CXI, Compiled by E. Parkes, London: HMSO, 1921.

Documents on British Foreign Policy 1919–1939, (ed. by) E. L. Woodward and R. Butler, First Series, London: HMSO, Vols. I–IV (1947–1956).

Documents on British Foreign Policy 1919–1939, (ed. by) R. Butler and J. P. T. Bury, First Series, London: HMSO, Vols. VII–VIII (1958) XII (1962) XIII (1963) XV (1967).

Documents on British Foreign Policy 1919–1939, (ed. by) W. N. Medlicot, D. Dakin and M. E. Lambert, First Series, London: HMSO, Vol. XVII (1970) XVIII (1972) XIX (1974) XX (1976) XXV (1984).

Great Britain, *Parliamentary Debates* (House of Commons) 1918–1923, vols. 109–68, London: HMSO, 1918–1923.

Great Britain, *Parliamentary Papers*, 1919, A Collection of Reports on Bolshevism, Command 8.

Great Britain, *Parliamentary Papers*, Army, August 1919, Expenditure on Military and Naval Operations in Russia from the Date of the Armistice to July 31, 1919, Command 307.

Great Britain, *Parliamentary Papers*, Army, November 1919, Expenditure on Military and Naval Operations in Russia from the Date of the Armistice to October 31, 1919, Command 395.

Great Britain *Parliamentary Papers*, 1920, Treaty Series No. 12, Command 963.

Great Britain, *Parliamentary Papers*, 1920, Treaty Series No. 11, Command 964.

Great Britain, *Parliamentary Papers*, 1921, Trade Agreement between His Britannic Majesty's Government and the Government of the Russian Socialist Federal Soviet Republic, Command 1217.

Great Britain, *Parliamentary Papers*, 1921, Report (Political and Economic) of the Committee to Collect Information on Russia, Command 1240.

Great Britain, *Parliamentary Papers*, 1923, Treaty Series, No. 16, Command 1299 (Lausanne).

Great Britain, *Parliamentary Papers*, 1923, Correspondence between His Majesty's Government and the Soviet Government Respecting the Relations between the Two Governments, Command 1869.

Great Britain, *Parliamentary Papers*, 1923, Reply of Soviet Government to His Majesty's Government Respecting the Relations Between the two Governments, Command 1874.

Great Britain *Parliamentary Papers*, 1923, Further Correspondence Between His Majesty's Government and the Soviet Government, Command 1890.

Great Britain, *Parliamentary Papers*, 1930, Treaty Series, No. 7, Command 3488.

Moberly, F. J. *History of the Great War Based on Official Documents, the Campaign in Mesopotamia 1914–18*, Compiled at the request of the Government of India under the direction of the Historical Section of the Committee of Imperial Defence, vol. IV, London: HMSO, 1923.

The Treatment of Armenians in the Ottoman Empire, Documents Presented to Viscount Grey of Fallodon, London, New York and Toronto: Hodder and Stoughton, 1916.

Greek Papers

Reply of the Hellenic Delegation to the Statements Submitted to the Peace Conference by the Bulgarian Delegation with Regard to the Policy of Bulgaria and its Claims to Thrace, (translated from French) New York: Oxford University Press American Branch, 1920.

The Greek White Book: Diplomatic Documents Relating to Italy's Aggression Against Greece, Washington D.C., 1943.

Venizelos, E. *Greece Before the Peace Congress of 1919*, (An English translation from the original) New York: Oxford University Press American Branch, 1919.

Soviet (and Comintern) Papers

Anglo–Sovetskie Otnosheniia 1921–1927 (Noty i Dokumenty) Moscow: Izdanie, Litizdata Narodnogo Komissariata po Inostrannim Delam, 1927.

Bor'ba za Ustanovlenie i Uprotsenie Sovietskoi Vlasti v Dagestane 1917–1921gg, Moscow: Izdatel'stvo Akademii Nauk SSSR, 1958.

Bulletin of the IVth Congress of the Communist International, Moscow, No. 17 (25 November 1922).

Bulletin of the IVth Congress of the Communist International, Moscow, No. 20 (28 November 1922).

Degras, J. *Communist International, 1919–1943: Documents*, Vol. I, London, New York, Toronto: Oxford University Press, 1956.

—— *Soviet Documents on Foreign Policy*, Vol. I, London, New York, Toronto: Oxford University Press, 1951.

Dokumenty i materialy po vneshnei politike zakavkaz'ia i gruzii, Tiflis, 1919.

Dokumenty Vneshnei Politiki SSSR, Moscow: Gosudarstvennoe Izdatel'stvo Politicheskoi Literatury, Vol. I (1957) II (1958) III (1959) IV (1960) V (1961) VI (1962).

Khronika, Revoliutsionnykh Sovety v Krimu 1917–1920 gg., (Partiinii Arkhiv Krimskogo Obkoma KP Ukrainii) Simferopol: Izdatel'stvo 'Krim', 1969.

Kommunisticheskii Internatsional, kratkii istoricheskii ocherk, Moscow: Izdatel'stvo Politicheskoi Literatury, 1969.

Kommunisticheskii Internatsional v Dokumentakh, (ed. by Bela Kun) Moscow: Partiinoe Izdatel'stvo, 1933.

Meijer, J. M. (ed.) *The Trotsky Papers, 1917–1922*, The Hague: Mouton and Co., Vol. I(1964).

Sovetsko–Germanskie Otnosheniia, 2 vols., Moscow: Izdatel'stvo Politicheskoi Literatury, 1968 and 1971.

Vos'maia Konferentsiia RKP (B) Protokoly, Moscow: Gosudarstvennoe Izdatel'stvo Politicheskoi Literatury, 1961.

Turkish Papers

Lozan Telgraflari [Turkish Diplomatic Documents on Lausanne] Ankara: TTK, I (1990).

TBMM Gizli Celse Zabitlari [Minutes of the Secret Sessions of the Turkish GNA] Ankara: Türkiye Is Bankasi Kültür Yayinlari, Vol. I (1985).

TBMM Zabit Ceridesi [Minutes of the Sessions of the Turkish GNA] 1920–1923, Vols. I–IX, XV, XVI, Ankara: TBMM Matbaasi, 1940–1961.

Türk Istiklal Harbi [Turkish War of Independence] 6 vols., Ankara: Genelkurmay Baskanligi Harb Tarihi Dairesi, 1962–1968.

Periodicals

Belleten, (1956–1973)
Birikim [Accumulation] (1976–1980 and 1990–1994)
Harp Tarihi Vesikalari Dergisi [Journal of War History Documents] (1952–1968)
Izvestiia, (1917–1923)
Kommunisticheskii Internatsional, XV–XXIII (1920–1923)
Krasnaia Nov, (1921)
Narodi Vostoka, (1920)
Novyi Vostok, (1922–1923)
Pravda, (1917–1923)
Tarih ve Toplum [History and Society] (1981–1993)
Tarih Vesikalari Dergisi [Journal of Historical Documents] Vol. I and II (1941) new series (1961)
The Communist (English Organ of the Comintern, published by the EC of the CPGB) (1921–1923)
The Times (1918–1923)
Voprosy Istorii (1945–1991)
Yakin Tarihimiz [Our Recent History] 4 Vols (1962–1963)
Yeni Cag [New Era] (Official Journal of the TKP)
Yeni Gün [New Day]
Ziya [The Light] (Turkish Journal of the BKP) (1920–1923)

Memoirs and Published Speeches

In English

Adivar, H. E. *Conflict of East and West in Turkey*, Lahore: Ashraf, 1935.
—— *Memoirs of Halide Edib*, New York: Century Co., 1926.
—— *The Turkish Ordeal*, London, 1928.
Aralov, S. I. 'In the Turkey of Atatürk (Reminiscences of an Ambassador)', *International Affairs*, No. 8 (August 1960) pp. 81–7; No. 10 (October 1960) pp. 97–103; No. 11(November 1960) pp. 96–102.
—— 'On Lenin's Instructions', *International Affairs*, No. 4 (April 1960) pp. 10–15.
Atatürk, M. K. *A Speech Delivered by Mustafa Kemal Atatürk. 1927*, Istanbul, 1963.
Baldwin, O. *Six Prisons and Two Revolutions*, London, 1924.
Child, R. W. *A Diplomat Looks at Europe*, New York: Duffield, 1925.
Churchill, Sir Winston *The World Crisis, Part IV, The Aftermath*, Library of Imperial History, 1974.
Dane, E. *British Campaigns in the Nearer East. 1914–1918*, Vol. II, London, New York and Toronto: Hodder and Stoughton, 1919.
De Robien, L. *The Diary of a Diplomat in Russia, 1917–1918*, Translated from the French by C. Sykes, New York: Praeger, 1970.
Djemal Pasha, *Memories of a Turkish Statesman 1913–1919*, London: Hutchinson and Co. Paternoster Row, 1922.

Dunsterville, L. C. *The Adventures of Dunsterforce*, London, 1932.

Grew, J. C. *Turbulent Era*, Boston: Houghton Mifflin, 1952.

Harding, S. 'The Case of Mrs Stan Harding', *The Nineteenth Century and After*, July 1922, pp. 1–16.

—— 'The Moscow Trial', *The Nineteenth Century and After*, August 1922, pp. 280–7.

Herron, G. D. *The Defeat in the Victory*, London: Cecil Palmer, 1921.

Hodgson, R. H. 'Memoirs of an Official Agent: Trading with Russia, 1921–1923', *History Today*, No. 4 (August 1954) pp. 522–8.

Krasin, Lyubov *Leonid Krassin: His Life and Work*, London, 1929.

Lenin, V. I. *Collected Works*, Moscow: Progress Publishers, Vols. XXVIII (1981) XIX (1980) XXX (1977) XXXI (1977) XXXII (1973) XXXIII (1976) XXXV (1980) XXXVI–XXXVII–XXXXII–XXXXIV (1977) XXXXV (1970).

—— *The National-Liberation Movement in the East*, Moscow: Foreign Languages Publishing House, 1957.

Liman von Sanders, O. *Five Years in Turkey* [Reprinted from the August Scherl ed., Berlin, 1920.] Annapolis, Md.: US Naval Institute, 1927.

Lloyd George, D. *Memoirs of the Peace Conference*, New Haven, vol. I (1938) vol. II (1939).

—— *The Truth about the Peace Treaties*, Vol. II, London: Victor Gollancz Ltd., 1938.

—— *War Memoirs of David Lloyd George*, Vols I and II, London: Odhams Press Limited, 1934–1936.

Lockhart, R. H. B. *Memoirs of a British Agent*, London: Macmillan, 1974.

Ludendorff, E. von *May War Memoirs 1914–1918*, 2 vols., London, 1919.

Luke, H. *Cities and Men*, Vol. II, London: Geoffrey Bless, 1953.

MacDonell, R. *And Nothing Long*, London: Constable and Co Ltd., 1938.

Maiskii, I. M. 'Diplomats of Lenin School: Georgi Chicherin', *New Times*, No. 44 (1 November 1967) pp. 10–13.

—— 'Diplomats of the Lenin School: Ambassador Krasin', *New Times*, No. 3 (24 January 1968) pp. 23–7.

Mayer, A. J. *Politics and Diplomacy of Peacemaking: Containment and Counterrevolution at Versailles, 1918–1919*, New York, 1967.

Miller, D. H. *My Diary at the Conference at Paris with Documents*, vol. III, New York, 1928.

Mikoyan, A. I. *Memoirs of Anastas Mikoyan*, I, Madison, Connecticut: Sphinx Press, 1988.

Nicolson, H. *Peacemaking, 1919*, London: Constable and Co Ltd, 1944 (first publication in 1933).

Price, G. W. *Extra-Special Correspondent*, London: George G. Harrap and Co. Ltd., 1957.

Rawlinson, A. *Adventures in the Near East, 1918–1922*, London: Jonathan Cape, 1934.

Riddell, Lord, *Lord Riddell's Intimate Diary of the Peace Conference and After, 1918–1923*, London: Gollancz, 1933.

Roy, M. N. M. *N. Roy's Memoirs*, Bombay: Allied Publishers Private Limited, 1964.
Seymour, C. *The Intimate Papers of Colonel House*, New York: Houghton Mifflin, 1928, vol. III.
Teague-Jones, *The Spy Who Disappeared*, London: Victor Gollancz Ltd., 1990.
Temperley, H. W. V. *A History of Peace Conference of Paris*, London: Oxford University Press, Vols. I, II, III (1920) VI and V (1921) VI (1924).
The Mist Procession. The Autobiography of Lord Vansittart, London, 1958.
The Papers of Woodrow Wilson, Princeton, N.J.: Princeton University Press, vol. XXXXV, 1984.
Zinoviev, G. 'Bolshevik Aims in Asia', *Current History*, XIII, Part 2, No. 3 (March 1921) p. 465.

In German

Ludendorff, E. von, *Meine Kriegserinnerungen*, Berlin, 1918.
Mühlmann, C. *Das Deutsch–Türkische Waffenbündnis im Weltkriege*, Leipzig: Verlag Koehler and Amelang, 1940.
Zinoviev, G. *Die russische Revolution und das internationale Proletariat*, Verlag der Kommunistischen Internationale, 1920.

In Russian

Aralov, S. I. *Vospominaniia Sovetskogo Diplomata*, Moscow: Izdatel'stvo Instituta Mezhdunarodnikh Otnoshenii, 1960.
Chicherin, G. V. 'Lozanskaia konferentsiia i mirovoe polozhenie', *Mezhdunarodnaia Zhizn*, No. 2 (1923) pp. 3–6.
—— 'Rossiia i aziatskie narody', *Vestnik NKID*, No. 2 (13 August 1919) pp. 1–7.
—— *Stat'i i rech'i po voprosam mezhdunarodnoi politiki*, Moscow: Sotsialno-Ekonomicheskoi Literaturi, 1961.
Denikin, A. I. *Ocherki russkoi smuty*, 5 vols., Paris and Berlin, 1921–1926.
Frunze, M. V. *Neizvestnoe i zabytoe*, Moscow: Nauka, 1991.
Kirov, S. M. *Rech'i i stat'i*, I, Moscow: Gosudarstvennoe Sots–Ekonomicheskoe Izd-vo, 1937.
Krasin, L. B. *Voprosy vneshnei torgovli*, Moscow and Leningrad, 1928.
Lenin, V. I. *Leninskii sbornik*, Nendeln, Liechtenstein: Kraus Reprint Ltd., 1966.
—— *Polnoe Sobranie Sochinenii*, Moscow: Gosudarstvennoe Izdatel'stvo Politicheskoi Literatury, Vols. XXXVI (1962) XXXVII–XXXXIII (1963) XXXXIV–XXXXV (1964).
Maiskii, I. M. 'Anglo–sovetskoe torgovoe soglashenie 1921 goda', V*oprosy Istorii*, No. 5 (May 1957) pp. 60–77.
Mikoyan, A. I. *Dorogoi Borbi*, I, Moscow: Izdatel'stvo Politicheskoi Literatury, 1971.

—— *Misli i Vospominaniia o Lenine*, Moscow: Izdatel'stvo Politicheskoi Literatury, 1970.

Narimanov, N. *Stat'i i pis'ma*, Moscow, 1925.

Ordzhonikidze, G. K. *Biografiia*, Moscow: Gosudarstvennoe Izdatel'stvo Politicheskoi Literatury, 1962.

—— *Izbrannye stat'i i rech'i, 1911–1937*, Moscow: Gosudarstvennoe Izdatel'stvo Politicheskoi Literatury, 1939.

Radek, K. *Vneshniaia Politika Sovietskoi Rossii*, Moscow and Petrograd, 1923.

Shaumian, S. G. *Izbrannye proizvedeniia*, Vol. II, Moscow, 1958.

—— *Stat'i i rech'i (1908–1918)* Baku: Kooperativnoe Izdatel'stvo 'Bakinskii Rabochii', 1924.

Stalin, J. *Sochineniia*, Moscow: Gospolizdat, 1947.

Trotsky, L. *Sochineniia*, Moscow and Leningrad: Gosizdat, 1926.

Vorovsky, V. V. *Stat'i i materialy po voprosam vneshnei politiki*, Moscow, 1959.

In Turkish

Afetinan, A. Atatürk Hakkinda Hatiralar ve Belgeler [Memoirs and documents concerning Atatürk] Ankara: Is Bankasi Yayinlari, 1959.

Akçuraoglu, Y. *Uç Tarz-i Siyaset*, Istanbul, 1938.

Aralov, S. I. *Bir Sovyet Diplomatinin Türkiye Hatiralari* [A Soviet Diplomat's Recollections on Turkey] Istanbul: Burçak Yayinlari, 1967.

Atatürk, M. K. *Nutuk* [Speech] Istanbul: Devlet Basimevi, 1938.

Atatürk'ün Söylev ve Demeçleri [The Speeches and Statements of Atatürk] Ankara: TTK Basimevi, 1961.

Atatürk'ün Tamim Telgraf ve Beyannameleri, 1917–1938 [Atatürk's instructions, telegrams and declarations] Ankara, 1964.

Atay, F. R. *Atatürk'ün Hatiralari 1914–1919* [Atatürk's recollections, 1914–1919] Ankara: Türkiye Is Bankasi Kültür Yayinlari, 1965.

Aydemir, S. S. *Makedonya'dan Orta Asya'ya Enver Pasa* [Enver Pasha from Macedonia to the Central Asia] Vol. III, 1914–1922, Istanbul: Remzi Kitabevi, 1972.

—— *Tek Adam* [Man Alone] 3 vols., Istanbul, 1963–5.

Cavit Bey *Idama Bes Kala* [Five to the Execution] Istanbul: Emre Yayinlari, 1993.

Cebesoy, A. F. *Milli Mücadele Hatiralari* [Recollections on the National Struggle] Istanbul: Vatan Nesriyat, 1953.

—— *Moskova Hatiralari* [Moscow Memoirs] Istanbul: Vatan Nesriyat, 1955.

Emre, A. C. '1920 Moskova'sinde Türk Komünistleri' [Turkish Communists in 1920's Moscow] *Tarih Dünyasi*, No. 2, January 1965.

Frunze'nin Türkiye Hatiralari [Recollections of Frunze on Turkey] Istanbul: Cem Yayinevi, 1978.

Igdemir, U. *Anafartalar Muhaberatina Ait Tarihçe* [Historical Account of the Battle of Anafartalar] Ankara, 1962.

Inönü'nün Söylev ve Demeçleri [The Speeches and Statements of Inönü] Istanbul: Milli Egitim Basimevi, 1946.

Karabekir, K. *Istiklal Harbimiz* [Our War of Independence] Istanbul, 1957.
Mustafa Subhi Kavgasi ve Düsünceleri [The Struggle and Ideas of Mustafa Subhi] Brussels (TKP publication) 1974.
Nadi, Y. *Cerkes Ethem Kuvvetlerinin Ihaneti* [Treachery of the forces of Ethem the Circassian] Istanbul, 1955.
Narimanov, N. *Seçilmish Eserleri* [Selected Works] Baku: Elm, 1973.
Okyar, F. *Uç Devirde Bir Adam* [One Man in Three Epochs] Istanbul, 1980.
Orbay, R. 'Rauf Orbay'in Hatiralari' [Memoirs of Rauf Orbay] *Yakin Tarihimiz*, Vol. I (1962).
Ozalp, K. *Milli Mücadele, 1919–1922* [National Struggle, 1919–1922] Vol. I, Ankara, 1971.
Simsir, B. N. *Ingiliz Belgelerinde Atatürk* [British Documents on Atatürk] Vols. III–IV, Ankara: TTK Basimevi, 1979–1983.
Sorgun, M. T. (ed.) *Bitmeyen Savas. Kutulamare Kahramani Halil Pasanin Anilari* [Unended Struggle. Memoirs of the Hero of Kutulamare, Halil Pasha] Istanbul, 1972.
Suphi, M. *Ilk Yazilar* [Early Writings] Istanbul: Amaç, 1989.
Tansu, Samih Nafiz, (narrated by Erturk, Husamettin) *Iki Devrin Perde Arkasi* [Background of Two Periods] Istanbul: Pinar Yayinlari, 1964.
Togan, Z. V. *Bugünkü Türkili (Türkistan ve Yakin Tarihi)* [Today's Turkic Country (Turkestan and Recent History)] Istanbul, Vol. I (1941–7).
Unüvar, V. *Istiklal Harbinde Bolseviklerle Sekiz Ay* [Eight Months with the Bolsheviks in the War of Independence] Istanbul: Hilmi Kitabevi, 1948.
Yalman, A. E. *Yakin Tarihte Gördüklerim ve Geçirdiklerim* [All I experienced in the recent history] Vol. I, Istanbul: Rey Yayinlari, 1970.

Secondary Works

In English

Abbot, G. T. *Turkey, Greece and the Great Powers*, London: Robert Scott, 1916.
Adelson, R. *Mark Sykes*, London: Jonathan Cape, 1975.
Agabekov, G. S. *OGPU, the Russian Secret Terror*, New York: Brentano's, 1931.
Ahmad, F. 'Politics and Islam in Modern Turkey', *Middle Eastern Studies*, (January 1991) pp. 3–21.
—— *The Making of Modern Turkey*, London and New York: Routledge, 1993.
—— 'The Political Economy of Kemalism', in Kazancigil, A. and Ozbudun, E. (eds) *Atatürk*, London, 1981.
Akbaba, I. 'The Peace Treaty of Lausanne', *Revue Internationale d'Histoire Militaire*, No. 46 (1980) pp. 234–8.
Alastos, D. *Venizelos*, London: Percy Lund Humphries and Co. Ltd., 1942.
Allen, W. E. D. and Muratoff, P. *Caucasian Battlefields: A History of the Wars on the Turco–Caucasian Border 1828–1921*, Cambridge: Cambridge University Press, 1953.
Anderson, M. S. *The Eastern Question 1774–1923*, London, 1966.

Andrew, C. 'The British Secret Service and Anglo–Soviet Relations in the 1920s. Part I: From the Trade Negotiations to the Zinoviev Letter', *The Historical Journal*, 20, 3 (1977) pp. 673–706.

Andrew, C. and Gordievsky, O. *KGB*, London: Hodder and Stoughton, 1990.

Armstrong, H. C. *Grey Wolf*, London: Barker, 1932.

Arslanian, A. H. 'Dunsterville's Adventures: A Reappraisal', *International Journal of Middle Eastern Studies*, XII, No. 2, September 1980.

Arslanian, A. H. and Nichols, R. l. 'Nationalism and the Russian Civil War. The Case of Volunteer Army–Armenian Relations, 1918–20', *Soviet Studies*, 31, No. 4 (1979) pp. 559–73.

Aspaturian, V. V. *Process and Power in Soviet Foreign Policy*, Boston: Little, Brown and Company, 1971.

Aston, G. 'The Freedom of the Straits', *The Nineteenth Century and After*, (November 1922) pp. 843–51.

Avalishvili, Z. *Independence of Georgia*, London, 1935.

Baker, R. S. *Woodrow Wilson and the World Settlement*, Vol. II, Garden City, 1922.

Beaman, A. H. 'Lausanne and Its Lessons', *The Nineteenth Century and After*, 93, No. DLIII (March 1923) pp. 318–27.

Beglar, N. 'The Constant Factor in Moscow's Turkish Policy', *Caucasian Review*, No. 1 (1955) pp. 12–21.

Bennigsen, A. 'Marxism or Pan–Islamism', *Central Asian Survey*, vol. 6, No. 2 (1987) pp. 55–66.

Bennigsen, A. and Wimbush, S. E. *Muslim National Communism in the Soviet Union: A Revolutionary Strategy for the Colonial World*, Chicago: University of Chicago Press, 1979.

Benvenuti, F. *The Bolsheviks and the Red Army, 1918–1922*, (translated from Italian) Cambridge University Press, 1982.

Beria, L. *On the History of the Bolshevik Organisations in Transcaucasia*, Moscow: Foreign Languages Publishing House, 1949.

Bierstadt, E. H. *The Great Betrayal*, London: Hutchinson and Co., 1924.

Bilsel, C. 'International Law in Turkey', *American Journal of International Law*, Vol. XXXVIII, No. 4 (October 1944).

Blank, S. 'Bolshevik Organizational Development in Early Soviet Transcaucasia: Autonomy vs. Centralization, 1918–1924', *Transcaucasia*, (ed. by R. G. Suny) The University of Michigan: An Arbor, 1983, pp. 305–34.

Brailsford, H. N. 'The Hope of Genoa', *New Republic*, No. 30 (26 April 1922) pp. 250–1.

Braunthal, J. *History of the International*, Vol. II (1914–1943) Bristol: Nelson, 1967.

Briggs, M. P. *George D. Herron and the European Settlement*, Stanford University Press, 1932.

Brinkley, G. A. *The Volunteer Army and Allied Intervention in South Russia, 1917–1921*, University of Notre Dame Press, 1966.

Buchan, J. (ed.) *The Baltic and Caucasian States*, London: Hodder and Stoughton Ltd., 1923.

Bunyan, J. *Intervention, Civil War and Communism in Russia, April–December 1918, Documents and Materials*, Baltimore: John Hopkins Press, 1936.

Busch, B. C. *Mudros to Lausanne: Britain's Frontier in West Asia, 1918–1923*, State University of New York Press, 1976.

Calwell, C. E. *Field–Marshall Sir Henry Wilson, Bart., G.C.B., D.S.O., His Life and Diaries*, London: Cassell, 1927, Vol. II.

Carr, E. H, *International Relations between the Two World Wars, (1919–1939)* London: Macmillan, 1965.

—— *A History of Soviet Russia, Vol. IV, The Interregnum, 1923–1924*, London: Macmillan, 1954.

—— *The Bolshevik Revolution 1917–1923*, III, London: Penguin Books, 1988 (first published in 1953).

—— *German–Soviet Relations Between The Two World Wars, 1919–1939*, Oxford University Press, 1952.

—— 'Radek's 'Political Salon' in Berlin 1919', *Soviet Studies*, No. 3 (April 1951) pp. 411–29.

—— *The Twenty Years' Crisis, 1919–1939: An Introduction to the Study of International Relations*, London: Macmillan, 1939.

Carrington, C. E. *The Life of Rudyard Kipling*, New York: Garden City, 1955.

Carswell, J. *The Exile: A Life of Ivy Litvinov*, London, Boston: Faber and Faber, 1983.

Castle, W. T. F. *Grand Turk*, London, New York, Melbourne: Hutchinson and Co., 1942.

Chabigi, V. E. 'Forty Years of Soviet Near Eastern Policy', *The East Turkish Review*, (April 1959) pp. 28–36.

Charachidze, G. 'The Georgian Communist Party and the National Question', *Caucasian Review*, No. 1 (1955) pp. 22–35.

Chossudovsky, E. M. 'Lenin and Chicherin: The Beginning of Soviet Foreign Policy and Diplomacy', *Millennium*, No. 3 (Spring 1974) pp. 1–16.

—— *Chicherin and the Evolution of Soviet Foreign Policy and Diplomacy*, Geneva: Graduate Institute of International Studies, 1973.

—— 'Genoa Revisited: Russia and Coexistence', *Foreign Affairs*, No. 50 (April 1971) pp. 554–77.

Churchill, Rogers P. *Anglo–Russian Convention of 1907*, Iowa: The Torch Press, 1939.

Clayton, A. *The British Empire As A Superpower, 1919–39*, London: Macmillan, 1986.

Cleveland, C. O. 'Memoirs of Chicherin', *Commonweal*, No. 24 (28 August 1936) pp. 417–18.

Coates, W. P. and Coates, Z. K. *A History of Anglo–Soviet Relations*, London: Lawrence and Wishart, 1943.

—— *Armed Intervention in Russia, 1918–1922*, London: Gollancz, 1935.

Conte, F. *Christian Rakovski (1873–1941)* New York, 1989.

Cosmetatos, S. P. P. *The Tragedy of Greece*, (translated by E. W. and A. Dickes) London: Kegan Paul, Trench, Trubner and Co. Ltd., 1928.

Craig, G. A. and Gilbert, F. (eds) *The Diplomats, 1919–1939*, Princeton, 1953.

Cumming, H. H. *Franco–British Rivalry in the Post-War Near East*, London: Oxford University Press, 1938.

Cunningham, Allan (ed.) *Early Correspondence of Richard Wood*, London, 1966.

—— *Eastern Question in the Nineteenth Century*, London: Frank Cass, 1993.

Dagobert, von M. *Mustapha Kemal*, London: William Heinemann, 1931.

Dakin, D. 'Lord Curzon's Policy Towards Greece', *Essays in Memory of Basil Laurdas*, Thessaloniki, 1975.

Darwin, J. G. 'The Chanak Crisis and the British Cabinet', *History*, Vol. 65, No. 213 (February 1980).

Davies, N. 'The Genesis of the Polish–Soviet War, 1919–1920', *European Studies Review*, No. 5 (January 1975) pp. 47–67.

—— *White Eagle, Red Star: The Polish–Soviet War, 1919–20*, New York: St. Martins Press, 1972.

Davis, K. W. *The Soviets at Geneva: The USSR and the League of Nations. 1919–1939*, Geneva: Librairie Kundig, 1934.

Davison, R. H. 'Middle East Nationalism: Lausanne Thirty Years After', *Middle East Journal*, 7, No. 3 (Summer 1953) pp. 326–44.

Debo, R. K. 'The Making of a Bolshevik: Georgii Chicherin in England 1914–1918', *Slavic Review*, No. 25 (December 1966) pp. 651–62.

Demirchian, K. S. *Soviet Armenia*, Moscow: Progress Publishers, 1984.

Diomidis–Petsalis, N. *Greece at the Paris Peace Conference 1919*, Thessaloniki, 1978.

Dontas, D. *Greece and Turkey*, Athens: G.C. Eleftheroudakis S.A., 1987.

Dukes, P. *Russia Under Catherine the Great*, Newtonville: Oriental Research Partners, 1978.

Dumont, P. 'French Freemasonry and the Turkish Struggle for Independence (1919–1923)', *International Journal of Turkish Studies*, III, No. 2 (1985–1986) pp. 1–15.

Dyer, G. 'The Turkish Armistice of 1918. 1. The Turkish Decision for a Separate Peace in Autumn, 1918', *Middle Eastern Studies*, No. 8 (1972) pp. 143–78.

—— 'The Turkish Armistice of 1918. 2. Lost Opportunity: the Negotiations of Moudros', *Middle Eastern Studies*, No. 8 (1972) pp. 313–48.

—— 'Turkish 'Falsifiers' and Armenian 'Deceivers': Historiography and the Armenian Massacres', *Middle Eastern Studies*, 12 (1976) pp. 99–107.

Earle, E. M. *Turkey, the Great Powers, and the Baghdad Railway*, New York: Macmillan, 1923.

Ellis, C. H. *The Transcaspian Episode 1918–1919*, London: Hutchinson, 1963.

Ellison, G. *An Englishwoman in Angora*, London: Hutchinson, 1923.

Essad–Bey *Secrets of the OGPU*, London: Jarrolds Publishers, 1933.

Etherton, Lt. Colonel P. T. *In the Heart of Asia*, London: Constable, 1925.

Evans, L. *United States Policy and the Partition of Turkey*, Baltimore: John Hopkins Press, 1965.

Eversley, Lord *The Turkish Empire*, London: T. Fisher Unwin Ltd., 1924.

Fair, J. 'The Peacemaking Exploits of Harold Temperley in the Balkans, 1918–1921', *Slavonic and East European Review*, 67, No. 1 (1989) pp. 68–93.

Fiddick, T. C. *Russia's Retreat from Poland, 1920*, London: Macmillan, 1990.

Finefrock, M. 'Atatürk, Lloyd George, and the Megali Idea: Cause and Consequence of the Greek Plan to Seize Constantinople from the Allies, June–August 1922', *Journal of Modern History*, 52, No. 1 (1980) pp. D1047–D1078.

Fink, C. *The Genoa Conference*, Chapel Hill and London: The University of North Carolina Press, 1984.

Fischer, L. *The Soviets in World Affairs: A History of the Relations Between the Soviet Union and the Rest of the World, 1917–1929*, Vol. I, Princeton University Press, 1951.

Fisher, H. A. L. 'Mr Lloyd George's Foreign Policy, 1918–1922', *Foreign Affairs*, No. 1 (15 March 1923) pp. 69–84.

Footman, D. *Civil War in Russia*, London: Faber, 1961.

Fraser, G. 'Basmachi-II', *Central Asian Survey*, Vol. 6, 2 (1987) pp. 7–42.

Frey, F. W. *The Turkish Political Elite*, Cambridge, Mass.: The MIT Press, 1965.

Gathorne-Hardy, G. M. *The Fourteen Points and the Treaty of Versailles*, Oxford: Clarendon Press, 1939.

Gawrych, G. 'Kemal Atatürk's Politico-Military Strategy in the Turkish war of Independence, 1919–1922: From Guerrilla Warfare to the Decisive Battle', *Journal of Strategic Studies*, 11, No. 3 (1988) pp. 318–41.

Gilbert, M. *Sir Horace Rumbold*, London: Heinemann, 1973.

—— *Winston S. Churchill*, London: Heinemann, Vol. IV (1916–1922) 1975.

Glasgow, G. 'Preliminaries of Genoa', *New Republic*, No. 30 (12 April 1922) pp. 193–6.

Glenny, M. V. 'The Anglo–Soviet Trade Agreement, March 1921', *Journal of Contemporary History*, Vol. 5/2 (April 1970) pp. 63–82.

Goldstein, E. 'Great Britain and Greater Greece 1917–1920', *Historical Journal*, 32, No. 2 (1989) pp. 339–56.

Gökalp, Z. *The Principles of Turkism*, Leiden: Brill, 1968.

Gökay, B. 'Turkish Settlement and the Caucasus, 1918–20', *Middle Eastern Studies*, 32, No. 2 (April 1996) pp. 45–76.

—— 'The Red Star and the Green Crescent', *Turkish Area Studies Group Newsletter*, No. 41 (November 1995) pp. 25–31

—— 'From Western Perceptions to Turkish Self-Perceptions, *Journal of Mediterranean Studies*, 5, No. 2 (1995) 259–69.

—— 'The Turkish Communist Party: Fate of the Founders', *Middle Eastern Studies*, 29, No. 2 (April–1993) pp. 220–35.

—— 'Historiography of the Post-War Turkish Settlement', *New Perspectives on Turkey*, No. 8 (Fall 1992) pp. 27–48.

—— 'Kinross on Atatürk', *Middle Eastern Studies*, 27, No. 4 (October 1991) pp. 694–99.

Grant, A. J. and Temperley, H. *Europe in the Nineteenth and Twentieth Centuries (1789–1950)* London–New York–Toronto: Longmans, Green and Co., 1952 (sixth revised edition).

Gratz, G. and Schuller, R. *The Economic Policy of Austria–Hungary during the War*, New Haven, 1928.

Graubard, S. R. *British Labour and the Russian Revolution, 1917–1924*, Cambridge: Harvard University Press, 1956.

Gregor, R. 'Lenin, Revolution and Foreign Policy', *International Journal*, No. 22 (Autumn 1967) pp. 563–75.

Gruber, H. *International Communism in the Era of Lenin: A Documentary History*, Ithaca: Cornell University Press, 1967.

Haithcox, J. P. *Communism and Nationalism in India*, Princeton University Press, 1971.

Hallgarten, G. W. F. 'General Hans von Seeckt and Russia, 1920–1922', *Journal of Modern History*, No. 21 (March 1949) pp. 28–34.

Hamilton, Sir I. *Gallipoli Diary*, New York: G. D. Doran, 1920.

Hancock, W. K. *Smuts: The Sanguine Years, 1870–1919*, Cambridge University Press, 1962.

Hankey, M. P. A. *The Supreme Command 1914–1918*, London, 1961.

Hardinge of Penhurst, Lord *The Old Diplomacy: The Reminiscences of Lord Hardinge of Penhurst*, London: John Murray, 1947.

Harington, Sir C. *Tim Harington Looks Back*, London: John Murray, 1940.

Harris, G. S. *The Origins of Communism in Turkey*, Stanford, California: Hoover Institution Press, 1967.

Haslip, J. *Catherine the Great*, London: Weidenfeld and Nicolson, 1977.

Hedlam, H. W. *The Black Sea, the Bosphorus and the Dardanelles. Studies in Diplomatic History*, London: Methuen, 1930.

Hehir, P. 'The Near East Crisis', *The Nineteenth Century and After*, DXLIX (November 1922) pp. 829–42.

Henderson, G. B. *Crimean War Diplomacy and Other Historical Essays*, Glasgow: Jackson, Son and Co., 1947.

Hilger, G. and Meyer, A. G. *The Incompatible Allies: A Memoir History of German–Soviet Relations 1918–1941*, New York, 1953.

Hodgson, J. E. *With Denikin's Armies. Being a Description of the Cossack Counter-Revolution in South Russia, 1918–1920*, London: Lincoln Williams, 1932.

Hoestler, C. W. *Turkism and the Soviets: The Turks of the World and Their Political Objectives*, London: Allen and Unwin, 1954.

Hooker, J. R. 'Curzon and the Curzon Line', *Journal of Modern History*, No. 30 (June 1958) pp. 137

Hopkirk, P. *On Secret Service East of Constantinople*, London: John Murray, 1994.

—— *The Great Game*, Oxford: Oxford University Press, 1990.

—— *Setting the East Ablaze*, Oxford: Oxford University Press, 1986.

Hopper, B. C. 'Narkomindel and Comintern: Instruments of World Revolution', *Foreign Affairs*, No. 19 (July 1941) pp. 737–50.

Housepian, M. *Smyrna, 1922: The Destruction of a City*, London, 1972.

Hovannisian, R. G. 'Armenia and the Caucasus in the Genesis of the Soviet–Turkish Entente', *International Journal of Middle Eastern Studies*, 4, No. 2 (1973) pp. 129–147.

—— 'The Armeno–Azerbaijani Conflict over Mountainous Karabagh, 1918–1919', *Armenian Review*, 24 (Summer 1971) pp. 3–24.

—— *Armenia on the Road of Independence, 1918*, Berkeley/Los Angeles, 1967.

—— 'The Allies and Armenia', *Armenian Review*, 17 (Winter 1964) pp. 20–39.
—— *The Republic of Armenia*, University of California Press, 1982.
Howard, H. N. *The Partition of Turkey*, Oklahama: University of Oklahama Press, 1931.
—— 'Turkish Foreign Policy', A*sia*, (January 1938) pp. 29–31.
Hurewitz, J. C. *Diplomacy in the Near and Middle East, A Documentary Record, 1535–1956*, Vols. I and II, Princeton, N.J.: D. Van Nostrand, 1956.
Isaacs, H. R. *The Tragedy of the Chinese Revolution*, (revised ed.) Stanford University Press, 1951.
Jelavich, B. *The Ottoman Empire, the Great Powers and the Straits Question, 1870–1887*, Bloomington, Ind., 1973.
Kahan, V. 'The Communist International, 1919–43: The Personnel of Its Highest Bodies', *International Review of Social History*, No. 21 (1976, Part 2) pp. 151–85.
Kapur, H. *Soviet Russia and Asia, 1917–1927. A Study of Soviet Policy towards Turkey, Iran and Afghanistan*, London: Michael Joseph, 1966.
—— *The Soviet Union and the Emerging Nations*, Geneva: Michael Joseph Ltd., 1972.
Karpat, K. H. *Ottoman Population, 1830–1914*, The University of Wisconsin Press, 1985.
Katkov, G. *The Kornilov Affair: Kerensky and the Break-up of the Russian Army*, London: Longmans, 1980.
Kazemzadeh, F. *The Struggle for Transcaucasia (1917–1921)* New York: Philosophical Library, 1951.
Kedourie, E. *England and the Middle East: The Destruction of the Ottoman Empire 1914–1921*, London: Bowes and Bowes, 1956.
—— *Politics in the Middle East*, Oxford, New York: Oxford University Press, 1992.
Kent, M. 'British Policy, International Diplomacy and the Turkish revolution', *International Journal of Turkish Studies*, III, No. 2 (1985–1986) pp. 33–51.
Kerr, S. E. *The Lions of Marash*, Albany, NY, 1973.
King, R. D. *Sergei Kirov and the Struggle for Soviet Power in the Terek Region, 1917–1918*, New York and London: Garland Publishing, Inc., 1987.
Kinnear, M. *The Fall of Lloyd George*, London and Basingstoke: Macmillan, 1973.
Kinross, Lord *Atatürk: A Bibliography of Mustafa Kemal, Father of Modern Turkey*, New York: William Morrow and Co., 1965.
Kitaigorodsky, P. 'The Labour Movement in Turkey', *The Communist International*, No. 12 (1925) pp. 83–96.
Knollys, D. E. 'Military Operations in Transcaspia, 1918–1919', *Journal of the Central Asian Society*, 13, Part. II (April 1926) pp. 89–110.
Kocharian, S. 'Soviet National Policy in Armenia', *Caucasian Review*, No. 1 (1955) pp. 81–92.
Kosok, P. 'Revolution and Sovietisation in the North Caucasus', *Caucasian Review*, No. 1 (1955) pp. 47–54.

Kousoulas, D. G. *Revolution and Defeat: The Story of the Greek Communist Party*, London, 1965.

Krassin [Krasin] L. 'The Anglo–Russian Trade Pact', *Living Age*, 23 April 1921, p. 212.

La Chesnais, P. G. *The Defence of the Cossacks Against Bolshevism*, Paris: Impr.-Libr. Militaire Universelle, 1919.

Ladas, S. P. *The Exchange of Minorities (Bulgaria, Greece and Turkey)* New York: Macmillan and Co., 1932.

Landau, J. M. *Pan–Turkism in Turkey: A Study of Irredentism*, London: Hurst, 1981.

Lazich, B. and Drachkovitch, M. M. *Bibliographical Dictionary of the Comintern*, California: Hoover Inst. Press, 1986.

—— *Lenin and the Comintern*, Vol. I, Stanford: Hoover Institution Press, Stanford University, 1972.

Leonidov, A. 'Socialism's First Diplomats', *New Times*, No. 28 (12 July 1967) pp. 12–16.

Lerner, W. *Karl Radek: The Last Internationalist*, Stanford: Stanford University Press, 1970.

—— 'Poland in 1920: A Case Study in Foreign-Policy Decision Making under Lenin', *South Atlantic Quarterly*, No. 72 (Summer 1973) pp. 406–14.

Lewis, B. *The Emergence of Modern Turkey*, London: Oxford University Press, 1968.

Lewis, G. *Modern Turkey*, New York: Praeger Publishers, 1974.

Lybyer, A. H. 'Turkey Under the Armistice', *Journal of International Relations*, XII, No. 4 (April 1922) pp. 456–8.

McCarthy, J. 'Greek Statistics on Ottoman Greek Population', *International Journal of Turkish Studies*, No. 2 (1980) pp. 66–76.

Macfie, A. L. 'The Chanak Affair (September–October, 1922)', *Balkan Studies*, 20, No. 2 (1979) pp. 309–41.

Marshall, D. *A Modern History of Georgia*, London: Weidenfeld and Nicolson, 1962.

Martin, T. S. 'The Urquhart Concession and Anglo–Soviet Relations, 1921–1922', *Jahrbucher für Geschichte Osteuropas*, No. 20 (December 1972) pp. 551–70.

Marx, K. *The Eastern Question, a Reprint of Letters Written 1853–56 Dealing with the Events of the Crimean War*, New York: Augustus M. Kelley Publishers, 1969.

Maurice, F. *The Life of General Lord Rawlinson of Trent from His Journals and Letters*, London: Cassell, 1928.

Mayer, A. J. *Politics and Diplomacy of Peacemaking: Containment and Counterrevolution at Versailles, 1918–1919*, New York: Knopf, 1967.

Mears, E. G. *Modern Turkey*, New York: The Macmillan Co, 1924.

Medvedev, R. *Let History Judge*, (translated from Russian by C. Taylor) New York: Alfred A. Knopf, 1971.

Menemencioglu, N. 'Enver Pasha in Turkestan, 1918–1922', *Revue Internationale d'Histoire Militaire*, No. 67 (1988) pp. 253–61.

Meyendorff, Baron A. 'My Cousin, Foreign Commissar Chicherin', *Russian Review*, No. 30 (April 1971) pp. 173–8.

Michalopoulos, C. 'Asia Minor and the Dodecanese', *Asiatic Review*, 16, No. 1 (January 1920) pp. 154–7.

Miller, A. E. 'The Origins of Leninist Eastern Policy', *International Affairs*, No. 4 (April 1972) pp. 68–75.

Mills, J. S. *The Genoa Conference*, New York: Dutton, 1922.

Mitrokhin, L. *Failure of Three Missions*, Moscow: Progress Publishers, 1987.

Modern Encyclopedia of Religions in Russia and the Soviet Union, vol. IV, Academic International Press, 1991.

Morgan, K. O. *Consensus and Disunity*, Oxford: Clarendon Press, 1979.

Mullaly, H. 'The Vital Eastern Problem', *The Nineteenth Century and After*, (December 1922) pp. 906–25.

Nalbandian, L. *The Armenian Revolutionary Movement*, University of California Press, 1965.

Nicolopoulos, J. 'The Testimony of Mikhail Vasilievich Frunze (1885–1925) concerning the Tragedy of Pontic Hellenism', *Journal of Modern History*, No. 4 (1987) pp. 37–53.

Nicolson, H. *George Curzon: The Last Phase, 1919–1925*, New York: Howard Fertig, 1974.

Nollau, G. and Wiehe, H. J. *Russia's South Flank*, London and Dunmow: Pall Mall Press, 1963.

Northedge, F. S. and Wells, A. *Britain and Soviet Communism*, London: Macmillan, 1982.

O'Connor, T. E. *The Engineer of Revolution: L. B. Krasin and the Bolsheviks, 1870–1926*, Boulder, Colo., 1992.

Orga, I. *Phoenix Ascendant: The Rise of Modern Turkey*, London: Robert Hale, 1958.

Paksoy, H. B. 'US and Bolshevik Relations with the TBMM Government: First Contacts. 1919–1921', *The Journal of Sophia Asian Studies*, XII (1994).

Palmer, A. W. *A Dictionary of Modern History*, Penguin Books, 1962.

Phillipson, C. and Buxton, N. *The Question of the Bosphorus and the Dardanelles*, London: Stevens and Haynes, 1917.

Pipes, R. *The Formation of the Soviet Union: Communism and Nationalism, 1917–1923*, Cambridge: Harvard University Press, 1964.

Ponomaryov, B., Gromyko, A. and Khvostok, V. (eds) *History of Soviet Foreign Policy, 1917–1945*, (English translation by David Skvirsky) Moscow, 1969.

Pope, A. U. *Maxim Litvinoff*, New York: Fischer, 1943.

Price, Clair *The Rebirth of Turkey*, New York, 1923.

Psomiades, H. *The Eastern Question: The Last Phase. A Study of Greek–Turkish Diplomacy*, Thessaloniki, 1968.

Radek, K. 'Lord Curzon and the Soviet Union', *Labour Monthly*, Vol. 7, No. 5, May 1925.

Rauf, B. The Last Sultans, Cheltenham, 1995.

Rigby, T. H. 'The Birth of the People's Commissars', *Australian Journal of Politics and History*, No. 20 (April 1974) pp. 70–5.

—— *Lenin's Government: Sovnarkom, 1917–1922*, Cambridge: Cambridge University Press, 1979.

Robinson, R. D. *The First Turkish Republic: A Case Study in National Development*, Cambridge, Mass.: Harvard University Press, 1963.

Ronaldshay, Earl of (Lawrence, J. L. D.) *The Life of Lord Curzon*, vol. III, London, 1928.

Rosenbaum, K. *Community of Faith*, Syracuse, New York, 1965.

Rothschild, J. *The Communist Party of Bulgaria: Origins and Development, 1883–1936*, New York: Columbia University Press, 1959.

Roy, M. N. 'Indian Revolutionaries in Moscow', *Radical Humanist*, 6 June 1954, pp. 270–1; and 13 June 1954, pp. 282–3.

Rudin, Harry R. *Armistice 1918*, New Haven: Yale University Press, 1944.

Rustow, D. A. 'The Army and the Founding of the Turkish Republic', *World Politics*, No. 11 (1949) pp. 513–52.

Quinault, R. 'Churchill and Russia', *War and Society*, vol. IX, 1 (May 1991) pp. 99–120.

Sachar, H. M. *The Emergence of the Middle East: 1914–1924*, New York: Alfred A. Knopf, 1969.

Scheffer, P. 'Maxim Litvinov: An Intimate Study', *Current History*, No. 34 (August 1931) pp. 670–77.

Senn, A. E. *Assassination in Switzerland: The Murder of Vatslav Vorovsky*, Madison: University of Wisconsin Press, 1981.

Serge, V. *From Lenin to Stalin*, Translated by R. Manheim, New York: Monad Press, 1973.

—— 'Litvinov', *Esprit*, No. 81 (June 1939) pp. 419–27.

Sforza, Count C. *Makers of Modern Europe: Portraits and Personal Impressions and recollections*, Indianapolis: Bobbs Merrill, 1928.

Shah, S. I. A. *Kamal: Maker of Modern Turkey*, London: Herbert Joseph, 1934.

Shapiro, L. (compiled and edited by) *Soviet Treaty Series*, 2 vols., Washington D.C.: The Georgetown University Press, 1950.

Shaw, S. J. and Shaw, E. K. *History of the Ottoman Empire and Modern Turkey*, Vol. II, Cambridge: Cambridge University Press, 1977.

Shotwell, J. *Turkey at the Straits*, London: Macmillan, 1940.

Shutleworth, D. I. 'Turkey, From the Armistice to the Peace', *Journal of Central Asian Society*, No. 11 (1924) pp. 61–2.

Smith, A. L. 'The German General Staff and Russia, 1919–1926', *Soviet Studies*, No. 8 (October 1956) pp. 125–33.

Smith, E. D. *Turkey: Origins of the Kemalist Movement and the Government of the Grand National Assembly*, Washington D.C.: Public Affairs Press, 1959.

Smith, M. L. *Ionian Vision*, London: Allen Lane, 1973.

Sonyel, S. R. 'Mustafa Kemal and the Turkish Nation's Struggle for Survival', *Cultura Turcica*, 4 (1967) pp. 70–198.

—— 'The Secret Anglo–Ottoman Treaty of 1919 in the Light of New Documents', *Belleten*, 34, No. 135 (July 1970) pp. 437–49.

—— *Turkish Diplomacy 1918–1923*, London: Sage Publications Ltd., 1975.

Sousa, N. *Capitulatory Regime of Turkey*, Baltimore: The John Hopkins Press, 1933.

Soviet Foreign Policy, I, (eds) A. A. Gromyko and B. N. Ponomarev, Moscow: Progress Publishers, 1981.

Soysal, I. 'Seventy Years of Turkish–Arab Relations and an Analysis of Turkish–Iraqi Relations (1920–1990)', *Studies on Turkish–Arab Relations*, No. 6(1991) pp. 23–84.

Stanwood, F. *War, Revolution and British Imperialism in Central Asia*, London: Ithaca Press, 1983.

Stavrianos, L. S. *The Balkans Since 1453*, Hinsdale, 1958.

Sumner-Boyd, H. and Freely, J. *Strolling Through Istanbul*, Istanbul: Redhouse Press, 1973.

Suny, R. G. *The Baku Commune, 1917–1918*, Princeton University Press, 1972.

Suvonov, V. *Soviet Military Intelligence*, London: Grafton, 1984.

Swietochowski, T. *Russian Azerbaijan, 1905–1920*, Cambridge: Cambridge University Press, 1985.

—— *Russia and Azerbaijan*, New York, 1995.

Sykes, C. *Wassmuss, 'The German Lawrence'*, London, New York, Toronto: Longmans, Green, 1936.

Szczesniak, B. (ed.) *The Russian Revolution and Religion*, Notre Dame, 1959, pp. 113–48.

Tamkoç, M. *A Bibliography on the Foreign Relations of the Republic of Turkey, 1919–1967, and Brief Biographies of Turkish Statesmen* , Ankara: METU Faculty of Administrative Sciences, No. 11, 1968.

Taylor, A. J. P. *The Struggle For Mastery in Europe, 1848–1918*, Oxford: Clarendon Press, 1954.

The Ottoman Empire in the Reign of Süleyman the Magnificent , I, Istanbul Research Centre, 1988.

Thompson, John M. *Russia, Bolshevism and the Versailles Peace*, Princeton University Press, 1966.

Thomson, Sir B. *The Allied Secret Service in Greece*, London: Hutchinson and Co. Ltd., 1931.

Toynbee, A. J. 'Angora and the East', *The Contemporary Review*, (June 1923) pp. 681–91.

—— *Armenian Atrocities: The Murder of A Nation*, London, New York and Toronto: Hodder & Stoughton, 1915.

—— 'The New Status of Turkey', *The Contemporary Review*, (March 1923) pp. 281–9.

—— *The Western Question in Greece and Turkey*, London: Constable, 1922.

Trotzky [Trotsky] L. *Lenin*, London, Calcutta, Sydney: George G. Harrap and Co. Ltd., 1925.

—— *Problems of the Chinese Revolution*, The University of Michigan Press, 1967.

Trukhanovsky, V. G. *Winston Churchill*, Moscow: Progress Publishers, 1978.

Trumpener, U. *Germany and the Ottoman Empire 1914–1918*, Princeton: Princeton Univ. Press, 1968.

Trush, M. I. ' A Diplomat of the Leninist School. For the 100th Anniversary of the Birth of G. V. Chicherin', *International Affairs*, No. 12 (December 1971) pp. 66–72.

Tschebotarioff, G. P. *Russia, My Native Land*, New York, London, Toronto: McGraw-Hill Book Company, 1964.

Tsoucalas, C. *The Greek Tragedy*, Penguin Books, 1969.

Uldricks, T. J. *Diplomacy and Ideology: The Origins of Soviet Foreign Relations 1917–1930*, Beverley Hills and London, 1979.

—— 'Russia and Europe: Diplomacy, Revolution, and Economic Development in the 1920s', *International History Review*, No. 1 (January 1979) pp. 55–83.

—— 'The Soviet Diplomatic Corps in the Chicherin Era', *Jahrbucher für Geschicte Osteuropas*, No. 3 (1975) pp. 213–14.

Ullman, R. *Anglo–Soviet Accord*, Princeton, NJ: Princeton University Press, 1972.

—— *Britain and the Russian Civil War*, Princeton, NJ: Princeton University Press, 1968.

—— *Intervention and the War*, Princeton, NJ: Princeton University Press, 1961.

Uratadze, G. 'The Suppression of an Element of Georgian Independence', *Caucasian Review*, No. 1 (1955) pp. 55–70.

Uregian, H. and Baghdjian, K. 'Two Unpublished Eyewitness Accounts of the Holocaust of Smyrna, September 1922', *Armenian Review*, 35, No. 4 (1982) pp. 362–89.

Vali, F. A. *Bridge Across the Bosphorus: The Foreign Policy of Turkey*, Baltimore: The John Hopkins Press, 1972.

Vat, Dan van der *The Ship That Changed The World*, London, Sydney, Auckland, Toronto: Hodder and Stoughton, 1985.

Vere-Hodge, *Turkish Foreign Policy 1918–1948*, Ambilly-Annemasse, 1950.

Von Mikusch, D. *Mustafa Kemal: Between Europe and Asia*, London: William Heinemann, 1931.

Wade, R. 'Why October? The Russian Search for Peace in 1917', *Soviet Studies*, No. 20 (July 1968) pp. 36–45.

Walder, D. *The Chanak Affair*, London: Hutchinson, 1969.

Ward, B. *Turkey*, Oxford University Press, 1942.

Webster, C. *The Foreign Policy of Palmerston*, Vol. II, London: G. Bell and Sons, Ltd., 1951.

Webster, D. E. *The Turkey of Atatürk*, Menasha, Wisc.: George Banta, 1939.

Wesson, R. G. *Soviet Foreign Policy in Perspective*, Homewood-Illinois: The Dorset Press, 1969.

Wheeler-Bennet, J. W. *Brest–Litovsk: The Forgotten Peace*, London: Macmillan, 1939.

White, S. 'Communism and the East: The Baku Congress, 1920', *Slavic Review*, No. 33 (September 1974) pp. 492–514.

—— 'Colonial Revolution and the Communist International, 1919–1924', *Science and Society*, No. 40 (Summer 1976) pp. 173–93.

—— 'Anti-Bolshevik Control Officers' and British Foreign Policy, 1918–1920', *Co-existence*, No. 13 (October 1976) pp. 144–56.

—— *Britain and the Bolshevik Revolution: A Study in the Politics of Diplomacy, 1920–1924*, London: Macmillan, 1979.
—— *The Origins of Détente: the Genoa Conference and Soviet–Western relations, 1921–1922*, Cambridge: Cambridge University Press, 1985.
Wise, E. F. 'Anglo–Russian Trade and the Trade Agreement', *Empire Review*, No. 28 (September 1923) pp. 995–1004.
Woolf, L. S. *The Future of Constantinople*, London: Unwin, 1917.
Wortham, H. E. *Mustafa Kemal of Turkey*, Boston: Little, Brown and Co., 1931.
Yapp, M. E. *The Making of the Modern Near East 1792–1923*, London and New York: Longman, 1987.
—— *The Near East Since the First World War*, London and New York: Longman, 1991.
Yate, A. C. 'The Near East', *Asiatic Review*, 19 (June 1923) pp. 19–20.
—— 'What Constantinople Means to Britain and Islam', *Asiatic Review*, 15 (July 1919) pp. 336–7.
Young, G. *Constantinople*, London: Methuen and Co. Ltd., 1926.
Zamir, M. 'Population Statistics of the Ottoman Empire in 1914 and 1919', *Middle Eastern Studies*, 17, No. 1 (1981) pp. 85–106.
Zapantis, A. L. *Greek–Soviet Relations, 1917–1941*, Boulder: East European Monographs, 1982.
Zarevand, *United and Independent Turania*, Leiden: E. J. Brill, 1971.
Zia Bey, Mufty-Zade *Speaking of the Turks*, London: Stanley Paul and Co., n.d.
Zürcher, E. J. *The Unionist Factor*, Leiden: E. J. Brill, 1984.
—— *Turkey*, London, New York: I. B. Tauris and Co Ltd., 1993.

In Russian

Agaev, Gb. 'Pis'mo Enver Pashe', *Kavkaz*, No. 9 (1937) pp. 25–9.
Airapetian, M. E. and Kabanov, P. *Leninskie printsipy vneshnei politiki*, Moscow: Politizdat, 1957.
Akhmedov, T. *Nariman Narimanov*, Baku: Yazichi, 1988.
Akhtamzian, A. A. 'Genuezskaia konferentsiia i Rapall'skii dogovor', *Voprosy Istorii*, No. 5 (May 1972) pp. 42–63.
Alikberov, G. A. *Pobeda sotsialisticheskoi revoluitsii v Dagestane*, Makhachkala: Dagknigoizdat, 1968.
Arkhangelskiy, Vl. *Frunze*, Moscow: Izdatel'stvo TsK. VLKSM, 1970.
Azerbaidzhanskaia SSR, Moscow: Gosudarstvennoe Izdatel'stvo Geograficheskoi Literatury, 1957.
Azizbekova, P.– Mnatsakanian, A.– Traskunov, M. *Sovetskaia Rossiia i Borba za Ustanovlenie i Uprochenie Vlasti Sovetov v Zakavkaze*, Baku: Azerbaidzhan Gosudarstvennoe Izdatel'stvo, 1969.
Bagirov, Y. A. *Iz istorii sovetsko–turetskikh otnoshenii v 1920–22 godakh*, Baku: Izd-vo A.N.Az.S.S.R., 1965.
Bakhov, A. S. *Na zare sovetskoi diplomatii. Organy sovetskoi diplomatii v 1917–1922 gg.*, Moscow: Mezhdunarodnye otnosheniia, 1966.
Berkhin, I. B. *Istoriia SSSR*, Moscow: Izdatel'stvo 'Vysshaia Shkola', 1966.

Bibliografiia Turtsii, 1917–1958, Moscow: Akademiia Nauk SSSR Institut Vostokovedeniia, 1959.

Bolshaia Sovetskaia Entsiklopediia, Vol. II (third publ. 1970) XI (third publ. 1973) XX (second publ. 1953) Moscow.

Bondarevskaia, L. G. 'Velikii Oktiabr i Krizis Kolonialnoi Politiki Velikobritanii Na Vostoka' in *Velikiy Oktiabr i Istoricheskii Progres*, (ed.) A. U. Chubarian, Moscow: Nauka, 1987.

Bor'ba za Uprochenie Sovetskoi Vlasti v Gruzii, Tiflis: Gosudarstvennoe Izdatel'stvo 'Sabchota Sakartvelo', 1959.

Chicherin, G. V. 'Lenin i vneshnaia politika', *Voprosy Istorii*, No. 3 (March 1957) pp. 20–5.

—— *Lenin i vneshnaia politika*, Moscow: Politizdat, 1977.

Chubarian, A. O. *Brestskii mir*, Moscow: Nauka, 1964.

—— *V. I. Lenin i formirovanie sovetskoi vneshnei politiki*, Moscow: Nauka, 1972.

Chulok, I. S. *Ocherki Istorii Batumskoi Kommunisticheskoi Organizatsii*, Izdatel'stvo 'Sabchota Adzhara', 1970.

Daniialov, G. D. *Stroitel'stvo Sotsializma v Dagestane, 1918–1937*, Moscow: Izdatel'stvo 'Nauka' Glavnaia Redaktsiia Vostochnoi Literatury, 1988.

Dantsig, B. *Turtsiia*, Moscow: Voennoe Izdatel'stvo Ministerstva Vooruzhennykh Sil Souza SSR, 1949.

Dokumenty po istorii Baku, Baku: Azerbaidzhanskoe Gosudarsvennoe Izdatel'stvo, 1978.

Dubner, A. *Bakinskii proletariat v gody revoliutsii*, Baku: Azgnin, 1931.

Frunze, M. V. *Voennaia i Politicheskaia Deiatelnost*, Moscow: Voennoe Izdatel'stvo, 1984.

Gadzhiev, A. *Iz Istorii Ovrazovaniia i Padeniia Iugo-zapadnoi Kavkazskoi (Karskoi) Demokraticheskoi Respubliki*, Baku: Elm, 1992.

Galoian, G. A. *Rabochee Dvizhenie i Natsionalnii Vopros v Zakavkaze 1900–1922*, Erevan: Izdatel'stvo, 'Aiastan', 1969.

Gareev, M. A. *M. V. Frunze - Voennii Teoretik*, Moscow: Voennoye Izdatel'stvo, 1985.

Garibdzhanian, G. B. *V. I. Lenin i Bolshevistickie Organizatsii Zakavkazia, (1893–1924)* Erevan: Armeniia SSSR Izdatel'stvo, 1967.

Gorokhov, I.- Zamiatin, L.- Zemskov, I. *G. V. Chicherin: Diplomat leninskoi shkoly*, Moscow: Politizdat, 1966.

Gugov, R. Kh. *Sovmestnaia bor'ba narodov Tereka za Sovetskuiu vlast*, Nal'chik: Kabardo-Balkarskoe Knizhnoe Izdatel'stvo, 1957.

Gurko-Kriazhin, V. A. 'Angliiskaia interventsiia v 1918–1919 gg. v Zakaspii i Zakavkaz'e', *Istorik-Marksist*, 2(1926) pp. 115–39.

—— 'Vozniknovenie natsionalno-osvoboditelnogo dvizheniia v Turtsii', *Novyi Vostok*, No. 23–4 (1928) pp. 268–75.

Guseinov, T. *Oktiabr' v Azerbaidzhane*, Baku: Azgiz, 1927.

Iskenderov, M. S. *Iz istorii bor'by Kommunisticheskoi Partii Azerbaidzhana za pobedu sovetskoi vlasti*, Baku: Azgosizdat, 1958.

Istoriia Kommunisticheskoi Partii Sovetskogo Soiuza, Moscow: Gosudarstvennoe Izdatel'stvo Politicheskoi Literatury, 1959.

Istoriia Vneshnei Politiki SSSR, Moscow: Izdatel'stvo 'Nauka', 1966.
Ivanov, V. M. *Marshall M. N. Tukhachevsky*, Moscow: Voennoe Izdatel'stvo, 1985.
Iz Istorii Grazhdanskoi Voiny v SSSR, Vol. III (February 1920–October 1922) Moscow: Izdatel'stvo 'Sovetskaia Rossiia', 1961.
Kabulski, S. 'Sovremennaia Turtsiia', *Mezhdunarodnaia Zhizn*, No. 3 (1925) pp. 25–44.
Kadishev, A. B. *Interventsiia i grazhdanskaia voina v Zakavkaz'e*, Moscow, 1960.
Karpova, R. F. *L. B. Krasin - sovetskii diplomat*, Moscow: Sotsekgiz, 1962.
Kharmandarian, S. V. *Lenin i stanovlenie Zakavkazskoi Federatsii, 1921–1923*, Erevan: Izdatel'stvo, 'Aiastan', 1969.
Kheifets, A. N. *Sovetskaia Diplomatiia i Narodni Vostoka 1921–1927*, Moscow: Nauka, 1968.
—— *Sovetskaia Rossiia i Sopredel'nye Strany Vostoka v Gody Grazhdanskoi Voiny 1918–20*, Moscow: Nauka, 1964.
Khinchuk, L. M. *K istorii anglo–sovetskikh otnoshenii*, Moscow: Gosudarstvennoe Izdatel'stvo, 1928.
Khovratovich, I. M. *Georgii Vasil'evich Chicherin*, Moscow: Mysl', 1980.
Khudadov, V. 'Sovremennyi Azerbaidzhan', *Novyi Vostok*, 3 and 4 (1923) pp. 156–61, 162–85.
Kocharli, F. K. *Nariman Narimanov*, Baku: Iz-vo A.N.Az.S.S.R., 1965.
Kommunisticheskaya Partiia Sovetskogo Soiuza, Vol. I (1898–1925) Moscow: Gosudarstvennoe Izdatel'stvo Politicheskoi Literatury, 1953.
Kondrashev, I. 'S. M. Kirov, Organizator bolshevistskogo pogpolya v tylu u Denikina', *Voprosy Istorii*, No. 7 (1947) pp. 3–23.
Kornev, H. *Litvinov*, Moscow: Molodaia gvardiia, 1936.
Kriazhin, V. 'Razdel Turtsii vo Vremia Mirovoi Voiny', *Novyi Vostok*, No. 4 (1923) pp. 49–57.
Kuznetsova, S. I. 'Krakh Turetskoi interventsii v Zakavkaze v 1920–1921 godakh', *Voprosy Istorii*, No. 9 (1951) pp. 143–56.
—— *Ustanovlenie sovetsko–turetskikh otnoshenii (K 40-letiiu Moskovskogo dogovora mezhdu RSFSR i Turtsiei)* Moscow: Izdatel'stvo vostochnoi literatury, 1961.
Lemin, I. M. *V. I. Lenin i voprosy vneshnei politiki SSSR*, Moscow: Znanie, 1959.
Liubimov, N. N. 'O V.V. Vorovskom', *Voprosy Istorii*, No. 10 (October 1971) pp. 138–43.
M. V. Frunze, Moscow: Voennoe Izdatel'stvo Ministerstva Oborony SSSR, 1965.
Maiski, I. M. 'Anglo–sovetskoe torgovoe soglashenie 1921 goda', *Voprosy istorii*, No. 5, May 1957
Malenkaia Entsiklopediia Velikaia Oktiabrskaia Sotsialisticheskaia Revolutsiia, Moscow: Izdatel'stvo 'Sovetskaia Entsiklopediia', 1968.
Mikoyan, A. I. *Dorogoi Borbi*, I, Moscow: Izdatel'stvo Politicheskoi Literatury, 1971.
—— *Misli i Vospominaniia o Lenine*, Moscow: Izdatel'stvo Politicheskoi Literatury, 1970.

Miller, A. E. 'Amerikanskii Plan Zakhvata Konstantinopolia i Prolivov v 1919', *Voprosy Istorii*, No. 3 (1951) pp. 61–79.

—— *Ocherki Noveishei Istorii Turtsii*, Moscow and Leningrad: Izdatel'stvo Akademii Nauk SSSR, 1948.

Mil'man, A. S. *Politicheskii stroi Azerbaidzhana v XIX-nachale XX vekov*, Baku: Azerneshir, 1966.

Mints, I. I. *Angliiskaia interventsiia i severnaia kontr-revoliutsiia*, Moscow: Gosudarstvennoe Sots.-Ek. Izdatel'stvo, 1931.

—— Trukhanovski, V. G.; Miller, A. F.;Samorukov, N. I.; Iakubovskaia, S. I. *Sovetskaia Rossiia i kapitalisticheskii mir v 1917–1923 gg.*, Moscow: Politizdat, 1957.

Nanitashvili, N. L. *Germanskii Kapital v Zakavkaz'e*, Tiflis: Izdatel'stvo Tbilisskogo Universiteta, 1982.

Ocherki Istorii Kommunisticheskikh Organizatsii Zakavkaz'ia, Tiflis: Izdatel'stvo TsK KP Gruzii, 1967.

Ocherki Istorii Kommunisticheskoi Partii Armenii, Erevan: Armianskoe Gosudarstvennoe Izdatel'stvo, 1964.

Ocherki Istorii Kommunisticheskoi Partii Gruzii 1883–1970, Tiflis: Izdatel'stvo TsK KP Gruzii, 1971.

Oznobishin, D. V. *Ot Bresta do Iur'eva. Iz istorii vneshnei politiki Sovetskoi vlasti 1917–1920 gg.*, Moscow: Nauka, 1966.

Patskhveriia, B. 'Mustafa Kemal Atatürk', *Aziia i Afrika Segodnia*, No. 12 (1963) p. 34.

Piyashev, N. F. (compiled by) *V. V. Vorovsky*, Moscow: Sotsial–Ekonomitsecki Izd., 1959.

Popov, N. N. *Natsionalnaia politika Sovetskoi vlasti*, Moscow: Izdatel'stvo 'Krasnaia Nov', 1924.

Rozaliev, Y. N. *V. I. Lenin i Turtsiia*, Moscow: 1960.

Saakian, R. G. *Franko–Turetskie Otnosheniia i Kilikiia 1918–1923*, Erevan: Izdatel'stvo an Armianskoi SSR, 1986.

Shamsutdinov, A. M. 'Natsional'no-osvoboditelnaya borba turetskogo narada v 1919–22 godov', *Sovetskoe Vostokovedenie*, No. 2 (1956) pp. 56–8.

Sheynis, Z. *Maksim Maksimovich Litvinov: Revolutsioner, Diplomat, Cheloviek*, Moscow: Izdatel'stvo Politicheskoi Literatury, 1989.

Shisko, L. *Rasskazy iz Russkoi Istorii*, Odessa: Izdanie Obedinenikh Film, 1917.

Shkarenkov, L. K. 'Belaia emigratsiia: Agoniia kontrrevoliutsii', *Voprosy Istorii*, No. 5 (1976) pp. 110–20.

Sorkin, G. Z. *Pervyi s'ezd narodov Vostoka*, Moscow, 1961.

Sovetskaia Vneshniaia Politika 1917–1945 gg., Moscow: 'Mezhdunarodnyi Otnosheniia', 1992.

SSSR i Turtsiia 1917–1979, Moscow: Glavnaia Redaktsiia Vostochnoi Literatury, 1981.

Stasova, E. D. *Vospominaniia*, Moscow: Sotsialno–Ekonomicheskoi Literatury, 1969.

Stein, B. 'Diplomatiia Antanty i proekt konferentsii na Printsevyikh Ostrovakh', *Voprosy Istorii*, No. 3 (1947) pp. 25–52.

—— 'Russkii Vopros' no Parizhskoi Mirnoi Konferentsii', *Voprosy Istorii*, No. 8 (1947) pp. 3–29.

—— *'Russkii vopros' v 1920–1921 gg.*, Moscow: Politizdat, 1958.

Sturua, D. G. 'Slavnii put borbi i pobeg', *Voprosy Istorii*, No. 5 (1957) pp. 57–67.

Sverdlov, A. Ia. *Grigorii Konstantinovich Ordhonikidze (Sergo)- Biografiia*, Moscow, 1962.

Trofimova, L. I. 'Stranitsa diplomaticheskoi deiatel'nosti G. V. Chicherina', *Voprosy Istorii*, No. 2 (February 1972) pp. 114–23.

Trush, M. I. 'Diplomat leninskoi shkoly (K 100-letiiu so dnia rozhdeniia G. V. Chicherina')', *Mezhdunarodnaia zhizn*, No. 11 (1972) pp. 88–96.

Turkan, G. A. (ed.) 'Doklad Ia. E. Rudzutaka o Genuezskoi konferentsii', *Istoricheski arkhiv*, No. 2 (March–April 1962) pp. 80–95.

Velikaia Oktiabr'skaia Sotsialisticheskaia Revoliutsiia Entsiklopediia, Moscow: Izdatel'stvo 'Sovetskaia Entsiklopediia', 1977.

Veltman, M. L.[M. Pavlovich] 'Kemalisticheskaia dvizheniev v Turtsii', *Krasnaia Nov*, No. 1 (1921) pp. 218–22.

—— 'Sovetskaia rossia i anglo-frantsuzskie intrigi na vostoke', *Kommunisticheskii Internatsional*, No. 14 (6 November 1920).

Vishnegradova, A. 'Revoliutsionnoe dvizhenie v persidskom Azerbaidzhane', *Novyi Vostok*, 2 (1922) pp. 249–55.

Voennaia Akademiia imeni M. V. Frunze, Moscow: Voennoe Izdatel'stvo Ministerstva Obopony SSSR, 1980.

Volkov, F. D. *Krakh angliiskoi politiki interventsii i diplomaticheskoi izoliatsii Sovetskogo gosudarstva (1917–1924 gg.)* Moscow: Politizdat, 1954.

Voprosy Istorii Narodov Kavkaze, Tiflis: 'Metsniereba', 1988.

Zarevand, *Turtsiia i panturanism*, Paris, 1930.

Zalkind, I. 'Iz pervykh mesiatsev Narodnogo Komissariata po Inostrannym Delam', *Mezhdunarodnaia zhizn*, No. 15 (7 November 1922) pp. 55–61.

Zarnitskii, S. V. and Sergeev, A. N. *Chicherin*, Moscow: Molodaia gvardiia, 1966.

—— and Trofimova, L. I. Sovetskoi *strany diplomat*, Moscow: Izdatel'stvo Politicheskoi Literatury, 1968.

Zhvaniia, G. K. *V. I. Lenin, TsK Partii, i Bol'sheviki Zakavkaz'ia*, Tiflis: Merani, 1969.

In Turkish

Agaoglu, S. *Babamdan Hatiralar* [Memoirs from my Father] Ankara: Zerbamat Basimevi, 1940.

—— *Kuvayi Milliye Ruhu* [The Spirit of the National Struggle] Istanbul: Nebioglu Yayinevi, 1945.

Aghamirov, M. M. A. *Birinci Rus Inkilabi Yillarinda Azerbaycan Ictimai Fikrinde Milli Meseleye Dair* [The National Question in the Intellectual Life of Azerbaijan during the Years of the First Russian Revolution] , Baku: Iz-vo A.N.Az.S.S.R., 1963.

Akar, A. *'Eski Tüfek' Sosyalistler* [Socialists from the old generation] Istanbul: Iletisim Yayinlari, 1989.

Akçuraoglu, Y. 'Türkçülük' [Turkism] *Türk Yili*, 1928, pp. 289–455.

Aksin, A. *Atatürk'ün Dis Politika Ilkeleri ve Diplomasisi* [Atatürk's Foreign Policy Principles and His Diplomacy] 2 Vols., Istanbul: Inkilap ve Aka Kitabevleri, 1962–1964.

'Amasya Mülakati Tutanaginin Tam Metni' [The Full Text of the Amasya Meeting Minutes] *BTTD*, No. 14 (November 1968) pp. 28–9.

Ariburnu, K. *Milli Mücadelede Istanbul Mitingleri* [The Istanbul Demonstrations during the War of Independence] Istanbul, 1951.

Askun, V. C. *Sivas Kongresi* [The Sivas Congress] Istanbul: Inkilap Kitabevi, 1963.

Ataöv, T. '1–7 Eylül 1920 Dogu Halklari Birinci Kongresinde Enver Pasa'nin Konusma Metni ve Bununla Ilgili Kongre karari' [The Speech of Enver Pasha in the Congress of Eastern Nationalities, 1–7 September 1920, and the Decision of the Congress on it] *Siyasal Bilgiler Facultesi Dergisis*, 29, No. 1–2 (1975) pp. 43–51.

Avcioglu, D. *Türkiye'nin Düzeni: Dün-Bugün-Yarin* [The Socio-Economic-Political Order of Turkey: Yesterday-Today-Tomorrow] Ankara: Bilgi Yayinevi, 1969.

Azerbaycan Kommünist Partiyasinin Tarihi [The History of the Azerbaijani Communist Party] Baku: K.P.Az. Institut Istorii Partii, Vol. I, 1958.

Azimov, H. *Azerbaycan Gazalarinda Sovyetler (1917–1918 li yillar)* [The Soviets in the Regions of Azerbaijan (Years of 1917–1918)] Baku: Elm, 1971.

Baykara, H. *Azerbaycan'da Yenilesme Hareketleri* [The Modernisation Movements in Azerbaijan] Ankara: Türk Kültürünü Arastirma Enstitüsü, 1966.

Bayur, H. *Türkiye Devletinin Hariciye Siyaseti* [The Foreign Policy of the Republic of Turkey] Istanbul: A. Sait Matbaasi, 1942.

—— 'Kuvay-i Milliye Devrinde Atatürk'ün Dis Siyasa Ile Ilgili Bazi Görüs ve Davranislari' [Some Ideas and Actions of Atatürk During the Period of National Struggle] *Belleten*, 20 (October 1956) pp. 659–83.

—— *Türk Inkilap Tarihi* [The History of the Turkish Revolution] Vol. I, books 1–2 (1963–1964); Vol. II, books 1–4 (1943–1952); Vol. III, books 1–3 (1953–1957) Ankara: TTK Basimevi.

—— *Atatürk: Hayati ve Eseri* [Atatürk: His Life and His Work] Ankara: Aydin Kitabevi, 1962.

—— 'Birinci Genel Savastan Sonra Yapilan Baris Antlasmalarimiz' [Our Peace Agreements Concluded After the End of the First World War] *Belleten*, 30, No. 117, Part II (January 1966) p. 149.

—— 'Azerbaycan Istiklali Drami' [The Tragic Story of the Independence of Azerbaijan] *Türk Kültürü*, No. 79 (May 1969) pp. 36–45.

—— *Azerbaycan Istiklal Tarihi* [The History of the Independence of Azerbaijan] Istanbul: Halk Yayinlari, 1975.

Birgen, M. 'Bizimkiler ve Azerbaycan' [Ours and Azerbaijan] *Yakin Tarihimiz*, 2, No. 18 (1962) pp. 157–8.

—— 'Türk Ordusunun Azerbaycana Yardimi' [The Help of the Turkish Army to Azerbaijan] *Yakin Tarihimiz*, 2, No. 5 (1962) pp. 44–5.

Biyiklioglu, T. *Trakya'da Milli Mücadele* [National Struggle in Thrace] Ankara, 1956.

—— *Atatürk Anadolu'da, 1919–1921* [Atatürk in Anatolia, 1919–1921] Ankara: TTK Basimevi, 1959.

—— 'Türk Istiklal Savasi Yillarinda Sovyet–Türk Dostlugu' [Soviet–Turkish Friendship During the Turkish Independence War] *Belleten*, 25, No. 99 (July 1961) pp. 486–91.

Borak, S. *Atatürk'ün Özel Mektuplari* [Atatürk's Private Letters] Istanbul: Varlik Yayinlari, 1970.

Bozdag, I. *Mustafa Suphi'yi Kim Oldürttü?* [Who Organised the Murder of Mustafa Subhi?] Istanbul: Emre Yayinlari, 1992.

Butbay, M. *Kafkasya Hatiralari* [Caucasian Recollections] Ankara: TTK Yayinlari, 1990.

Büyük Oktobir ve Azerbaycanda Sovyet Hakimiyeti Ugrunda Mubariza [Great October and the Struggle for Soviet Victory in Azerbaijan] Baku: Izd-vo A.N.Az.S.S.R., 1958.

Caferoglu, A. *Azerbaycan*, Istanbul: Cumhuriyet Matbaasi, 1940.

Criss, B. *Isgal Altinda Istanbul* [Istanbul Under Occupation] Istanbul: Iletisim, 1993.

Cerkez Ethem'in Hatiralari [Memoirs of Cerkez Ethem] Istanbul: Dünya M., 1962.

Culhaoglu, M. 'Cerkez Etem Olayi ve Tarih Bilmeceleri' [The Cerkez Ethem Affair and Historical Questions] *Gelenek*, No. 3 (January 1987) pp. 95–114.

Duran, I. 'Dünya Savasi Sonunda Türk Diplomasisinin Ilk Basarisi: Brest–Litovsk Hazirliklari' [First Successes of the Turkish diplomacy at the End of the First World War: Preparations for the Brest–Litovsk Meeting], *Belgelerle Türk Tarihi Dergisi*, April–June 1973.

Dursunoglu, C. *Erzurum Kongresi Sirasinda Atatürk'ün Düsünceleri* [Atatürk's Views During the Erzurum Congress] Ankara: TTK Basimevi, 1967.

—— *Milli Mücadelede Erzurum* [Erzurum During the National Struggle] Ankara: Ziraat Bankasi Matbaasi, 1946.

Emre, A. C. '1920 Moskova'sinda Türk Komünistleri' [The Turkish Communists in 1920s Moscow] *Tarih Dünyasi*, No. 1–3 (December 1964–February 1965) pp. 146–51 and 278–85.

Erdogan, F. *Türk Illerinde Hatiralarim* [Recollections in the Turkish Lands] Istanbul: Yeni M., 1954.

Erer, T. *Enver Pasa'nin Türkistan Kurtulus Savasi*,[Enver Pasha and the Civil War in Turkestan] Istanbul, 1971.

Erkin, F. C. *Türk–Sovyet Iliskileri ve Bogazlar Meselesi* [The Turco–Soviet Relations and the Question of the Turkish Straits] Ankara: Basnur Matbaasi, 1968.

Eroglu, H. *Türk Devrimi Tarihi* [The History of the Turkish Revolution] Ankara: Kardes Matbaasi, 1967.

Gologlu, M. *Erzurum Kongresi* [Erzurum Congress] Ankara: Nüve-Basnur M., 1971.

Gökay, B. 'Ankara'da Bir Ingiliz Casusu' [A British Spy in Ankara] *Tarih ve Toplum*, February 1992, no. 98.

—— 'Yunan Komünistleri Anadolu'da' [Greek Communists in Anatolia] *Tarih Toplum*, July 1992.

Gönlübol, M. and Sar, C. *Atatürk ve Türkiye'nin Dis Politikasi, 1919–1938* [Atatürk and Turkey's Foreign Policy, 1919–1939] Istanbul: Milli Egitim Basimevi, 1963.

Gövsa, I. A. *Türk Meshurlari Ansiklopedisi* [The Encyclopedia of Turkish Notables] Istanbul: Yedigün Nesriyati, 1946.

Hanioglu, M. S. *Osmanli Ittihad ve Terakki Cemiyeti ve Jön Türklük* [Ottoman Committee of Union and Progress and the Young Turks] Istanbul: Iletisim Yayinlari, 1985.

Hasircioglu, T. 'Kazim Karabekir, Mustafa Kemal Pasayi Tutuklamayi Reddediyor' [The Refusal of Kazim Karabekir Pasha to Arrest Mustafa Kemal] *BTTD*, 4 (January 1968) pp. 11–14.

Hayit, B. 'Sovyet Kaynaklarinda Atatürk' [Atatürk in the Soviet sources] *Belgelerle Türk Tarihi Dergisi*, No. 1 (March 1985) pp. 20–5.

Ileri, R. N. *Atatürk ve Komünizm* [Atatürk and Communism] Istanbul: Anadolu Yayinlari, 1970.

—— *Türkiye Komünist Partisi Gerçegi ve Bilimsellik - Quo Vadis Ibrahim Topçuoglu?* [The Reality of the Communist Party of Turkey and the Scientific Approach - Quo Vadis Ibrahim Topçuoglu?] Istanbul: Anadolu Yayinlari, 1976.

Jaeschke, G. 'Mondros'a Giden Yol' [The Road to Mudros] *Belleten*, 28, No. 109 (January 1964) pp. 141–52.

—— 'Ingiliz Belgelerinin Isigi Altinda Yunanlilarin Izmir Cikartmasi' [The Greek Occupation of Izmir in the Light of British Documents] *Belleten*, 32, No. 128 (October 1968) pp. 567–76.

—— *Türk Kurtulus Savasi Kronolojisi* [Chronology of the Turkish War of Independence] vol. I, Ankara, 1970.

Kandemir, F. *Atatürk'ün Kurdurdugu TKP* [The TKP Founded on the Instructions of Atatürk]Istanbul: Yakin Tarihimiz Yayinlari, 1965.

—— *Mustafa Kemal, Arkadaslari ve Karsisindakiler* [Mustafa Kemal, his Friends and his Opponents] Istanbul, 1964.

Karal, E. Z. *Mehmed VI. Muhammed VI Vahid al-Din*, IA, vol. 7, 1955.

—— *Osmanli Tarihi* [Ottoman History] Vols V, VI, VII, Ankara: TTK Basimevi, 1988.

Karaman, S. S. *Istiklal Mücadelesi ve Enver Pasha - Trabzon ve Kars Hatiralari: 1921–1922* [National Struggle and Enver Pasha - Trabzon and Kars Memoirs: 1921–1922] Izmir: Selüloz Basimevi, 1949.

Keykurun, N. *Azerbaycan Istiklal Mücadelesinin Hatiralari* [Recollections on the Struggle for the Independence of Azerbaijan] Istanbul: Azerbaycan Gençlik Dernegi, 1964.

Kiliç Ali Hatiralarini Anlatiyor [Kiliç Ali Narrates his Memoirs] Istanbul: Hisar M., 1955.

Kuran, A. B. *Inkilap Tarihimiz ve Ittihat Terakki* [Our History of Revolution and the Union and Progress] Istanbul: Tan M., 1948.

—— *Osmanli Imparatorlugunda ve Türkiye Cumhuriyetinde Inkilap Hareketleri* [Revolutionary Movements in the Ottoman Empire and the Turkish Republic] Istanbul: Celtüt M., 1959.

Kurat, A. N. *Türkiye ve Rusya: XVIII Yüzyil Sonundan Kurtulus Savasina Kadar Türk–Rus Iliskileri* [Turkey and Russia: Turco–Russian Relations from the End of the 18th century to the National Struggle] Ankara: Ankara Universitesi Yayinlari, 1970.

Kurtulan, E. 'Türk Ordusunun Azerbaycan'da Kalmasi Meselesi' [The Issue of the Turkish Army's Remaining in Azerbaijan] *Azerbaycan*, 4–6 (1963) pp. 13–17.

Kutay, C. *Atatürk-Enver Pasa Hadiseleri* [Atatürk-Enver Pasha Incidents] Istanbul, 1956.

—— *Milli Mücadelede Yesil Ordu Efsanesi* [The Green Army Legend in the National Struggle] Istanbul: Ercan M., 1956.

—— *Türkiye'de Ilk Komünistler* [Early Communists in Turkey] Istanbul: Ercan M., 1956.

—— *Cerkez Ethem Dosyasi* [The Dossier of Ethem the Circassian] Istanbul: Bogaziçi Yayinlari, 1977.

Küçük, Y. 'Solun Tarihine Baslangiç' [Beginning of the History of the Left] *Ekin Belleten*, Summer 1988, pp. 37–56.

—— *Türkiye Uzerine Tezler 1908–1978* [Theses on Turkey 1908–1978] Istanbul: Tekin Yayinlari, 1978.

Mazici, N. *Atatürk Döneminde Muhalefet* [Opposition in the Time of Atatürk] Istanbul: Dilmen Kitabevi, 1984.

Menekse, N. *Bulgar Komünist Partisinin Gizli Raporu: Türkiye'de Komünizm* [The Secret Report of the Bulgarian Communist Party: Communism in Turkey] Ankara: Ayyildiz M., 1967.

Meram, A. K. *Türk–Rus Iliskileri Tarihi* [History of the Turco–Russian Relations] Istanbul: Kitas Yayinlari, 1969.

Mirza-Bala, M. *Azerbaycan Misak-i Milli: 28 Mayis Istiklal Beyannamesinin Tahlili* [Azerbaijani National Pact: Analysis of the 28 May Declaration] Istanbul: Istiklal Matbaasi, 1927.

Mumcu, U. *Kazim Karabekir Anlatiyor* [Kazim Karabekir Narrates] Istanbul: Tekin Yayinevi, 1990.

Mustafa Subhi Kavgasi ve Düsünceleri [The Struggle and Ideas of Mustafa Subhi] Brussels (TKP Publication) 1974.

Müderrissoglu, A. *Kurtulus Savasinin Mali Kaynaklari* [Financial Sources of the War of Independence] Ankara: Maliye Bakanligi, 1974.

Nesimi, A. *Türkiye Komünist Partisi'nde Anilar / Degerlendirmeler* [Memoirs on the Communist Party of Turkey / Discussions] Istanbul: Promete, 1979.

Olaylarla Türk Dis Politikasi, 1919–1965 [The Foreign Policy of Turkey through Events, 1919–1965] (Compiled by M. Gönlübol, C. Sar, A. S. Esmer, O.

Sander, H. Ulman, A. S. Bilge, D. Sezer.) Ankara: Disisleri Bakanligi Matbaasi, 1968.

Ozerdim, S. N. *Atatürk Devrimi Kronolojisi* [Chronology of the Atatürk Revolutions] Ankara, 1974.

Rüstü, K. 'Büyük Harpte Baku Yollarinda 5-nci Piyade Firkasi' [5th Regiment on the Road to Baku During the Great War] *Askeri Mecmua*, (1934) pp. 1–49.

Sayilgan, A. *Pan–Türkçü Sosyalizmin Kaynaklari ve Sultan Galiev* [The Sources of the Pan–Turkic Socialism and Sultan Galiev] n.d.

—— *Solun 94 Yili (1871–1965: Türkiye'de Sosyalist–Komünist Hareketler)* [94 Years of the Left: (1871–1965: Socialist and Communist Movements in Turkey] Ankara: Mars M., 1968.

Selek, *Anadolu Ihtilali* [The Anatolian Revolution] Istanbul: Burçak Yayinlari, 1966.

Sertel, Z. M. *Hatirladiklarim 1905–1950* [Those I Remember] Istanbul: Yaylacik M., 1968.

Sertoglu, M. 'Amasya Protokolünün Tam ve Gerçek Metni' [The fFull and Real Text of the Amasya Declaration] *Belgelerle Türk Tarihi Dergisi*, No. 3 (December 1967) pp. 9–13.

—— 'Trabzon Bölgesinde Rum Pontus Cumhuriyeti Kurulmasi Faaliyeti' [The Activities Surrounding the Foundation of the Greek Pontus Republic in the Region of Trabzon] *Belgelerle Türk Tarihi Dergisi*, No. 11 (August 1968) pp. 3–7.

—— 'Anadolumuz Hakkinda Yunanlilarin Düsünce ve Emelleri' [The projects and aims of the Greeks on our Anatolia] *Belgelerle Türk Tarihi Dergisi*, No. 19 (April 1969) pp. 14–19.

Sirel, M. 'Sevr Antlasmasini Kabul Eden Saltanat Surasi' [the Assembly of the Sultanate which Accepted the Treaty of Sevres] *Belgelerle Türk Tarihi Dergisi*, No. 4 (January 1968) pp. 15–21.

Sonyel, S. R. '1919 Yili Ingiliz Belgelerinin Isiginda Mustafa Kemal ve Milli Mukavemet' [Mustafa Kemal and the national resistance movement in 1919 under the light of the British documents] *Türk Kültürü*, No. 85 (November 1969, pp. 47–8.

—— 'Yunan Millet Meclisi Gizli Tutanaklarinda Mustafa Kemal ve Türk Kurtulus Savasi' [Mustafa Kemal and the Turkish War of Independence in the secret minutes of the Greek National Assembly] *Belgelerle Türk Tarihi Dergisi*, No. 49(October 1971), pp. 11–19.

—— 'Yunan Millet Meclisi Gizli Tutanaklarinda Mustafa Kemal ve Türk Kurtulus Savasi' [Mustafa Kemal and the Turkish War of Independence in the secret minutes of the Greek National Assembly] *Belgelerle Türk Tarihi Dergisi*, No. 49 (October 1971) pp. 11–19.

—— 'Orgeneral Kazim Ozalp'in Anilari ile Ilgili Bir Açiklama' [An explanation about the memoirs of Brig.-General Kazim Ozalp] *Belleten*, 36, No. 146 (April 1973) pp. 231–4.

Swietochowski, T. *Müslüman Cemaatten Ulusal Kimlige Rus Azerbeycani, 1905–1920* [Russian Azerbaijan, from a Muslim community to the national identity, 1905–1920] Istanbul: Baglam Yayinlari, 1988.

Sehidoglu, S. H. *Eylül 1920– Baku Kurultayi ve Milli Mücadeleye Sagdan Soldan El Koyma Cabalari* [September 1920–Baku Congress and the attempts of the Left to control the national struggle] Istanbul: Ofset 75, 1975.

Sismanof, D. *Türkiye'de Isçi ve Sosyalist Hareketi* [The labour and socialist movements in Turkey] Istanbul: Belge Yayinlari, 1978.

Tanör, B. *Türkiye'de Yerel Kongre Iktidarlari (1918–1920* [The Governments of the Local Congresses in Turkey] Istanbul: AFA Yayincilik, 1992.

Tansel, S. *Atatürk ve Kurtulus Savasi 1919–1922* [Atatürk and the War of Independence 1919–1922] Ankara: Ajanstürk M., 1965.

Taymas, A. (A. Battal) *Rus Ihtilalinden Hatiralar–I, 1917–1919* [Memoirs on the Russian Revolution-I, 1917–1919] Istanbul: Güven B., 1949.

Tengirsek, Y. K. 'Milli Mücadelede Ruslarla Ilk Temaslarimiz' [Our Early Contacts with the Russians During the National Struggle] *Yakin Tarihimiz*, 4, no. 43 (20 December 1962).

—— *Vatan Hizmetinde* [On the Service of the Country] Istanbul, 1967.

Tevetoglu, F. 'Mustafa Kemal Pasa-General Harbord Görüsmesi' [Mustafa Kemal Pasha-General Harbord Meeting] *Türk Kültürü*, No. 7 (March 1969).

—— *Türkiye'de Sosyalist ve Komünist Faaliyetler, (1910–1960)* [Socialist and Communist Activities in Turkey, (1910–1960)] Ankara, 1967.

TKP Dogusu, Kurulusu, Gelisme Yollari [The birth of the TKP, its foundation, and the ways for its development] (TKP publication, no date).

Topçuoglu, I. *Neden Iki Sosyalist Parti 1946 – TKP Kurulusu ve Mücadelesinin Tarihi 1914–1960* [Why Two Socialist Parties 1946–the foundation and struggle of the TKP] Istanbul: Eser M., 1976.

Tunaya, T. Z. *Devrim hareketleri Içinde Atatürk ve Atatürkçülük - Müsahedeler ve Tezler* [Atatürk and Atatürkism within the Revolutionary Movements - Observations and Theses] Istanbul: Baha M., 1964.

—— *Türkiye'de Siyasi Partiler, 1859–1952* [Political Parties in Turkey, 1859–1952] Istanbul: Dogan Kardes, 1952.

Tunçay, M. *1923 Amele Birligi* [1923 Labour Union] Istanbul: BDS, 1989.

—— *Eski Sol Uzerine Yeni Bilgiler* [New Information on the Old Left] Istanbul: Belge Yayinlari, 1982.

—— 'Türk–Sovyet Iliskileri Tarihinde: Atatürk'le Ilgili Olarak Uydurulmus Bir Hikaye' [An unfounded claim about Atatürk in the history of Turco–Soviet relations] , *1973 Sinan Yilligi*, Istanbul: Sümer M., 1973.

—— *Türkiye'de Sol Akimlar* [Left-wing movements in Turkey] Ankara: Bilgi Yayinevi, 1967.

Turan, S. *'Kurtulus Savasinda Kongreler', Cumhuriyetin 50. Yil Semineri* ['The Congresses in the War of Independence', The Seminar for the 50th year of the Republic] Ankara: TTK Yayinlari, 1975.

Türk Tarih [Turkish History] Ankara: Devlet Basimevi, 1932 (vol. IV).

Türkgeldi, A. *Mondros ve Mudanya Mütarekelerinin Tarihi* [The History of the Armistices of Mudros and Mudania] Ankara: Güney Matbaacilik, 1948.

Ulug, I. H. *Atatürk'ün Cizdigi Portreler* [Pictures Drawn by Atatürk] Istanbul: Inkilap ve Aka, 1968.

Umar, B. *Izmir'de Yunanlilarin Son Günleri* [The Last Days of the Greeks in Izmir] Ankara: Bilgi Yayinlari, 1974.

Uzunçarsili, I. H. *Osmanli Tarihi* [Ottoman History] Vol. IV, Books 1–2, Ankara: TTK Basimevi, 1988.

Velikov, S. *Kemalist Ihtilal ve Bulgaristan, 1918–1922* [The Kemalist Revolution and Bulgaria] Istanbul: Kitas Yayinlari, 1969.

Yerasimos, S. *Azgelismislik Sürecinde Türkiye* [Turkey in the process of underdevelopment] III, Istanbul: Gözlem Yayinlari, 1976.

—— *Türk–Sovyet Iliskileri* [Turco–Soviet Relations] Istanbul: Gözlem Yayinlari, 1979.

Index